Edible Insects
Edited by Heimo Mikkola

Published in London, United Kingdom

IntechOpen

Supporting open minds since 2005

Edible Insects
http://dx.doi.org/10.5772/intechopen.77835
Edited by Heimo Mikkola

Contributors
Amritpal Singh Kaleka, Gaganpreet Kour Bali, Navkiran Kaur, Felix Meutchieye, C M (Tilly) Collins, Flora Dickie, Monami Miyamoto, Jacob Paarechuga Anankware, Michael Feldman, Rafael Cartay, Vladimir Dimitrov, Harry McDade, Heimo Juhani Mikkola

Notice
Statements and opinions expressed in the chapters are these of the individual contributors and not necessarily those of the editors or publisher. No responsibility is accepted for the accuracy of information contained in the published chapters. The publisher assumes no responsibility for any damage or injury to persons or property arising out of the use of any materials, instructions, methods or ideas contained in the book.

First published in London, United Kingdom, 2020 by IntechOpen
IntechOpen is the global imprint of INTECHOPEN LIMITED, registered in England and Wales, registration number: 11086078, 7th floor, 10 Lower Thames Street, London, EC3R 6AF, United Kingdom
Printed in Croatia

British Library Cataloguing-in-Publication Data
A catalogue record for this book is available from the British Library

Additional hard and PDF copies can be obtained from orders@intechopen.com

Edible Insects
Edited by Heimo Mikkola
p. cm.
Print ISBN 978-1-78985-635-4
Online ISBN 978-1-78985-636-1
eBook (PDF) ISBN 978-1-83968-451-7

We are IntechOpen,
the world's leading publisher of
Open Access books
Built by scientists, for scientists

4,600+
Open access books available

119,000+
International authors and editors

135M+
Downloads

151
Countries delivered to

Our authors are among the
Top 1%
most cited scientists

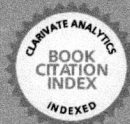

12.2%
Contributors from top 500 universities

Interested in publishing with us?
Contact book.department@intechopen.com

Numbers displayed above are based on latest data collected.
For more information visit www.intechopen.com

Meet the editor

Heimo Mikkola has a PhD and is an adjunct professor from the University of Kuopio (now part of the Eastern Finland University). He has worked for over 30 years in Africa, South America, and Central and Southeast Asia, mainly with the Food and Agriculture Organization of the United Nations (FAO). As the resident representative of FAO in Africa and South America, he familiarized himself with edible insects as a future possibility to feed the ever-increasing population of the world. After retiring from FAO, he has been a part-time professor in three Kazakh universities and one Kyrgyz university. He is lecturing on green biotechnology and global food security , which includes insect farming. He has published almost 650 work reports and scientific papers, mainly on applied zoology. In 2017, he edited an IntechOpen book on 'Future Foods' which contains five chapters on edible insects.

Contents

Preface

Edible insect farming is one of the most promising solutions for food production and food security in the future world when we may have nine billion or more people to feed [1]. Therefore, the growth potential of insect consumption is currently attracting a lot of interest at a global scale, thereby justifying this new book. The recent Future Foods book [1] included five chapters on edible insects.

Edible insects are a reliable source of fats, many minerals, and micronutrients. For more than a decade we have known that insects have a wide variety of fat- and water-soluble vitamins, making them a valuable food source for people with specific health needs. Insects as a food group have been found to be healthier that some meat alternatives [2].

High intakes of red meat and poultry are associated with increasing economic development but have costly environmental impacts in terms of greenhouse gasses, ammonia emissions, water use, and land area needed [3]. A case study in Hungary found that the production of edible insects might attract consumers seeking new food options and who intend to reduce meat intake [4].

Feed conversion efficiency in insect farming is four times better than that of normal livestock farming. Insect farming is socially inclusive as they are easy to farm and require no high-tech skills or large investments. Insect farming could help reduce the use of antibiotics in feed for chicken, fish, and pigs. Insect meal can replace 25–100% of soymeal or fish meal in the feed of our domesticated animals or fish. The aroma, taste, and carcass quality is not affected when using the insect meal, and overall levels of essential amino acids are very good as is the protein digestibility [5].

Some 3,000 indigenous population groups in over 110 countries in Asia, Australia, and Central and South America are consuming, almost daily, up to 2,000 distinct species of insects [6]. However, insect eating is declining in many traditional markets and is often seen as taboo in western cultures. Its recent introduction has been met by a range of barriers. Nutritional arguments are not thought to be enough to overcome the 'disgust factor' and convert westerners to insect-based dishes [7]. This book reviews several of these barriers and seeks new concepts in promoting the wider acceptance of insect eating.

Two years ago, the legislation, or rather the lack of it, was the largest barrier to developing insect production and marketing in Europe but now the legislation is no more a problem. However, some insect food producers in Finland concluded in 2019 that the marketing of insects was better when they were sold illegally as human food. But surely, the new legislation provides necessary advice and standards in appropriate production methods and for marketing of the hygienic insect products?

I wish to thank the Publishing Process Manager Lada Božić for her active efforts to assist in getting this book ready. Without her positive attitude many important chapters and necessary corrections would not have been published.

Heimo Mikkola
Eastern Finland University,
Finland

References

[1] Mikkola H. Preface in the Future Foods. InTech. 2017; VII-X. http://dx.doi.org/10.5772/65132

[2] Oonincx DGAB, Dierenfeld ES. An investigation into the chemical composition of alternative invertebrate prey. Zoo Biology. 2012;**31**(1):40-54. DOI: 10.1002/zoo.20382

[3] Collins CM, Vaskou P, Kountouris Y. Insect food products in the western world: Assessing the potential of a new 'Green' market. Annals of the Entomological Society of America. 2019;**112**(6):518-528. https://doi.org/10.1093/aesa/saz015

[4] Gere A, Székely G, Kovács S, Kókai Z, Sipos L. Readiness to adopt insects in Hungary: A case study. Food Quality and Preference. 2017;**59**:81-86. https://doi.org/10.1016/j.foodqual.2017.02.005

[5] Vantomme P. World population growth and the search for new protein alternatives. Conference Paper 2015. Available from: https://researchgate.net/publication/282753885

[6] Hardouin J. Production d'insectes à des fins économiques ou alimentaires: Mini-élevage et BEDIM. Notes Fauniques de Gembloux. 2003;**50**:15-25

[7] Deroy O, Reade B, Spencer C. The insectivore's dilemma, and how to take West out of it. Food Quality and Preference. 2015;**44**:44-55. https://doi.org/10.1016/j.foodqual.2015.02.007

Chapter 1

Introductory Chapter: Is the Insect Food Boom over or when it Will Start?

Heimo Mikkola

1. Introduction

As an introductory chapter for this "Edible Insects" book, I have collected a number of newspaper articles from Finland (Finnish names of the articles translated in English) on the insect food business development between 2016 and 2019. These papers show at first enormous boom of the insect food production and sales for human consumption after that business was finally legalized in September 2017. However, already in late 2018, many insect farmers and market people expressed their concerns that the business development has not been as lucrative as anticipated. Year 2019 brought more not so positive evaluations of insect food markets in Finland but concluding that the insect food boom will come one day, latest with the next generation of people.

2. First illegal food due to European Union regulations

Insect food marketing was at first illegal in Finland although some production existed. Enthusiastic farmers and researchers started, however, in 2016 to organize insect food tasting events in some restaurants and schools around the country. One restaurant in Helsinki (Restaurant Rupla, Helsinginkatu 16) had a three course insect menu one evening every two month (21.07, 21.09 etc.). None of the participants refused to eat the insect food [1]. Actually many people had to queue to get to taste the insect food.

In February 2017, newspapers wrote that many start-up companies in Otaniemi university campus believe that insect food business could bring millions income to the participants [2]. A bit later it was written that people in North Karelia are well prepared to start insect food production and that 70% of the Finnish population is ready to taste insect food products [3].

Finally in September 2017, Finland adapted the EU regulations so that it was acceptable to start selling insect food as human food products. This sales permit allows selling insect products made out of the following insects:

Black soldier fly larvae	*Hermetia illucens*
Cricket	*Acheta domesticus*
Desert locust	*Schistocerca gregaria*
Drone bee larvae	*Apis mellifera*
Lesser mealworm (= chicken hog larva)	*Alphitobius diaperinus*
Migratory locust	*Locusta migratoria*

1Tropical domestic cricket	*Gryllodes sigillatus*
Yellow mealworm	*Tenebrio molitor*

Several farmers modified their farms by giving up the pig farming and starting to grow insects, instead. Some tens of restaurants started to offer insect food on their menu. Many new insect products were introduced to the market when the sale of insect food became legal. Some producers stated that "We will start slowly, learning the markets and markets getting to know us." When asked about the taste, people stated: "Taste is mild, something between chicken and shrimps. Best crickets are when well fried" [4].

Leader Foods Oy, for instance, started to sell cricket protein bars. One Zircca bar contains 15 crickets and 34% protein and it is gluten free product. The company says that cricket bars meet all nutrition requirements equally well as any meat or fish products. And the iron content of crickets is higher than that in spinach [5].

On June 11, 2018, Oy Halva Ab brought to the markets cricket liquorice bars first time in Finland and in the world [6].

One of the most amazing problems in cricket farms in Finland have been extremely warm and dry summer weathers we have had recent years. Even the tropical species stopped eating due to the heath or rather due to the dry air. In the tropics, weather is always humid even in high temperatures. So the farmers were forced to invest in expensive air humidifiers [7]. Second major problem many commercial insect farmers faced when starting large scale production was the lack of commercial feed for the insects [8]. In 2018, the Natural Resources Institute Finland and Eastern Finland University started a project to produce plant-based pelleted feed for the insect farmers. This 'Hyvä Rehu' (=Good Feed) project lasts 2 years, and is funded by the Ministry of Agriculture of Finland.

At the end of 2018, newspapers started to write negative news from the farmers who had invested money to make millions with the insect food [9]. In Loviisa town, there was a huge insect farm that aimed to be the largest in Europe but had to be closed down in 2019 as not profitable enough [7]. After the boom started in 2017, the sales have gone down and supermarkets have started to diminish the selection of insect food products. Some insect food producers say that marketing of insects was better when they were sold illegally as human food. The products were often labeled and sold as kitchen and food decoration items [10].

Large part of the population still finds the insect food too exotic, but the producers and sale people hope that the next generation would be a real insect-eating generation.

Similarly large supermarket chains, S-group & K-markets, will keep the insect food available hoping that the sales will eventually pick up. When that will happen is still unknown [10].

One of the largest insect food companies in Finland is Finsect which also exports the insect products under the name "Griidy". They have 26 contract farmers mainly in Western Finland. They are producing cricket bread, cricket chocolate, cricket liquorice, roasted and seasoned crickets and cricket meal 150–450 E/kg for the consumers depending on the package size [11]. This high consumer price may partly explain the insect food marketing problems in Finland.

Additional papers collected but not cited

Kurki E. Pests in flour bags soon legal food table items. Karjalainen. 2016
Rouvinen M. Many Finns would like to buy insect food already now. Karjalainen. 2017

Savolainen S. Would you like to eat crickets? Apu. 2017

Merimaa J. Food security from insects. Helsingin Sanomat. 2017

Lehtinen T. Now insects will be made into food products. Helsingin Sanomat. 2017

Salminen J. Crickets feel in mouth like the Finnish Rye Crispbread. Helsingin Sanomat. 2017

Nieminen K. Insects can now be sold as food products. Karjalainen. 2017

Massinen T. Crickets for hunger: Insects scurrying to the plate/insect food boom is only starting. Karjalainen. 2019

Author details

Heimo Mikkola
University of Eastern Finland, Finland

*Address all correspondence to: heimomikkola@yahoo.co.uk

IntechOpen

References

[1] Tarvonen H-M. Insects creeping to the plates. Helsingin Sanomat. 2016;**30**:06

[2] Moilanen K. Cricket beef to the food table. Helsingin Sanomat. 2017;**04**:02

[3] Kurki E. Ackwardness is only a vision error: North Karelia is ready to start production of insect food. Karjalainen. 2017;**16**:02

[4] Kojo H. Insect food: Crickets are already chirping. Karjalainen. 2017;**17**:08

[5] Leader Foods Oy. Crickets conquerred the protein bars. Apu 6/2017

[6] Oy Halva Ab. Available at: https://www.halva.fi [Accessed 17/09/2019]

[7] Liiten M. Growing insects for food ended before starting. Helsingin Sanomat. 2019;**30**:01

[8] Koistinen A. Luke (=natural resources institute Finland) and university (= eastern Finland university) developing feed for the insects. Karjalainen. 2018;**10**:01

[9] Latvala J. Time was not ripe for the insect food, yet. Karjalainen. 2018;**23**:12

[10] Tapio K. Large part of the population finds that insect food products are yet too exotic. Karjalainen. 2018;**23**:12

[11] Finsect. Antti Reen. Helsingin Sanomat. 2019

Chapter 2

How Might We Overcome 'Western' Resistance to Eating Insects?

Harry McDade and C. Matilda Collins

Abstract

Entomophagy, the consumption of insects as a food source, occurs at a global scale with over 2 billion people seeing it as traditional. This practice does not extend into mainstream Western culture where its introduction is often met by a range of barriers, leaving entomophagy often seen as a taboo. The 'disgust response' of food neophobia and a lack of social and cultural contexts that reduce adoption may be overcome by strategic application of tools arising from innovation diffusion theory: relative advantage; compatibility; low complexity; trialability and observability. This chapter accessibly reviews known barriers to uptake and outlines the potential application of these concepts in promoting the wider acceptance of entomophagy.

Keywords: neophobia, taboo, acceptance, innovation, experience

1. Introduction

The growth potential of entomophagy is currently attracting much interest [1, 2]. Currently, this practice is declining in many traditional markets and does not extend into mainstream western culture where its introduction is often met by a range of barriers, leaving insect consumption often seen as a taboo [3–5]. Insect protein has great potential to be used as reliable alternative or supplement to vertebrate 'meat' consumption and offers relative advantages over traditional animal protein sources if entry barriers can be overcome. One advantage is the lower environmental impact of mass rearing insects in terms of greenhouse gasses and ammonia [6]. Furthermore, insects are highly nutritious and have been found to be healthier than some meat alternatives [7].

This chapter accessibly reviews known barriers to uptake and uses Rogers *Diffusion of innovation* theory (**Figure 1**) to outline possible strategies to overcome these [8].

2. Disgust and food neophobia

Insects can trigger a disgust response for a number of reasons. Disgust is different to an innate distaste reaction, which is a response to the bitterness of many

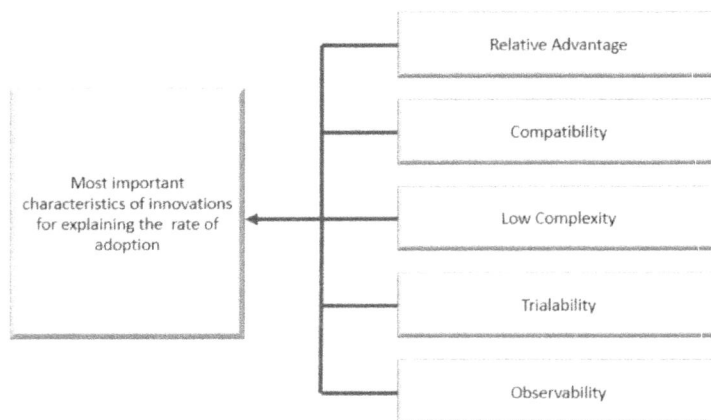

Figure 1.
Perceived characteristics of an innovation considered to determine the level of penetration into a target audience: relative advantage—the level to which the innovation is perceived to be better than existing alternatives; compatibility—the degree to which the innovation is perceived to be in keeping with values and experiences of the target population; complexity—the level to which an innovation is perceived as difficult to utilise and understand; trialability—the level that a new innovation can be experimented with; observability—the level to which the outcomes of an innovation is viewable by the target population [8].

biologically toxic compounds [9]. Rather than being a reflex action, a disgust reaction comes from a cognitive process when assessing foods and explains why differences are seen in cultural perception of entomophagy [10]. Disgust can arise with perceived or real associations of insects to objects of core disgust, which include pathogens and pathogen-related stimuli such as faecal matter and vomit [11]. Scaled disgust ratings can predict participants' willingness to attend an event with insect-based cuisine; this demonstrated clearly that disgust is a barrier to introduction of entomophagy into western diets [12].

Separate from the disgust response is the effect of food neophobia which also contributes to an unwillingness to try entomophagy. Food neophobia is simply the tendency to avoid the consumption of novel foods and the degree of novelty correlates strongly with willingness to try unfamiliar foods [13, 14].

2.1 Disgust and the 'law of contamination'

The law of contamination states that a disgust reaction will be elicited not only by objects of core disgust but through any objects that have been contacted. Rozin's elegant demonstration of this with fruit juice and sterile cockroaches indicates that this barrier to entomophagy is often based on irrational thought. Participants knew their reaction was irrational due to the cockroach having been sterilised [15, 16]. If the reaction does occur even with the knowledge that the organism is safe, then providing information on safety is likely to have little effect on uptake.

Overcoming this barrier to introduction may involve new brands initially selling insect-related products to focus primarily on gaining consumer trust or using established and 'trusted' brands to reduce the perceived risk of novel products [17]. Discovering and adopting shared values with consumers may permit entomophagy to become more compatible with western consumption while reducing negative attitudes [18]. Although, for this strategy to find success, individual brands must avoid perceived or real negative impacts on any group of consumers as these would likely impact the entomophagy industry as a whole by reducing trust [19]. For such trust to grow, a foundation of legislation is developing to provide advice and standards in appropriate methodology for hygienic insect handling and storage [20, 21].

2.2 Disgust response to 'identifiable insects'

Disgust also arises when consumers are reminded that they are eating an animal or are made aware of the animals' origins [11, 12]. When whole insects are found within a food product, this is considered an extended example of the law of contamination, as it occurs due to an association with dead animals and decay [11]. Many studies have pointed to invisibility of insects (such as in cookies) leading to increased willingness compared to their unprocessed counterparts (such as mealworms and crickets) [7, 10, 22]. These support the idea that, for easier implementation, innovators should focus on products without visible insects and thus simplify the product's trajectory to western acceptability.

2.3 Food neophobia

Insect protein products are seen as novel, which influences consumer perception and thus their expected experience on trying it [23]. Increased familiarity reduces any anticipated negative assumptions of taste and experience before trying them [24] and incorporating novel food into familiar dishes will accelerate consumer acceptance. This plays into Rogers' concepts of **compatibility** with western society as well as that of **low complexity**. The latter in this case is achieved through individuals being familiar with how common dishes in their culture are created and consumed. Mimicking familiarity also plays a part and people are more willing to try an insect when it comes in the format of a familiar food item [25]. Expectation also plays a role, an expectation of good flavour was found to be an important indicator for willingness to eat for crickets and silkworms [10].

There are examples where food neophobia has been overcome effectively. Rationing of U.S food supplies during World War II promoted consumption of unfamiliar organ protein. A strategy of preparing and serving these novel ingredients in expected and visually familiar ways led to accelerated uptake [24]. This may, however, prove more difficult with insects as they are perceptually distinct from mainstream food products. In order to capitalise on **observability**, using novel foods in a side dish accompanying a highly favoured familiar main dish can reduce variation in specific perception and in overall evaluation of the meal [26]. Thus, introducing insect protein to side dishes with the 'delicious' main course could optimise their introduction to mainstream diets before incorporation into main dishes.

Making a dish familiar is not enough, it is still important for the product to actually be a strongly positive culinary experience in terms of taste and texture. An *a priori* negative perception may become justified if the dish displays textural characteristics that the consumer does not enjoy and then serve to reinforce or increase aversion to insect dishes [13, 27]. Investment in the gastronomic integrity of dishes as well as in enticing advertising messages will increase the success of insect trials and encourage repeat consumption leading to increased **observability** [16]. There is much positive feedback between brand and item in the context of gaining consumer trust.

3. Social context and current culture

3.1 Absence of social context

Western culture has little recent experience with entomophagy and this is a barrier to its introduction as diet aligns strongly with the social norms of immediately surrounding culture [28]. This lack of social context for entomophagy allows for a greater level of food neophobia as all insect-based cuisine is seen as a novel food. Harnessing social norms may prove to be a method of increasing insect consumption as almost one-third of participants in one study tried insects 'in company' having first stated they would not [29]. This study concluded that having positive social models could result in mitigation of the disgust response. Expanding entomophagy as a social norm through positive models for people to **observe** and **trial** for themselves would thus increase **compatibility**.

3.2 Receptivity and age

Introducing children to entomophagy may bring these social norms into the general populace. People who tried foods in early childhood, even on rare occasions, were more likely to enjoy those foods when they were older. Parental influence is a less reliable indicator of liking foods when older, though this can encourage initial consumption of insect protein [30]. Social influence can be incorporated into the strategy by having parents and teachers as a positive model; **observing** adult influencers consuming insect products may draw greater willingness to try from the children. The challenge is how to develop the adult model to suit the most receptive 'primary school' age range [7]. Introduction to children should incorporate both visual and taste exposure to insect products; however, the focus should be on providing taste exposure to children as this has been shown to increase preference for the food item to a greater extent [31].

3.3 Complexity through absence of social context

Lack of social norms and context also increases the complexity of accepting entomophagy as innovation. With little opportunity for observability, people are less aware of the options available for entomophagy, where to begin, or even if it is possible to adopt it into their lifestyle. Creating social context is vital in allowing individuals to observe entomophagy before trying, it shows them that such dietary options are possible and can be desirable. The approach sometimes taken is that of 'bug banquets', events that offer the chance for consumers to try insect products. This approach can be biased as those people who seek out these experiences are more likely to have more positive views on entomophagy or lower neophobia scores. Furthermore, while these often result in reduced aversion to entomophagy, there is little to no follow-up on whether there is long-term uptake [32].

An alternative strategy for creating social context is to use social media. Applications such as Instagram, which has a high presence of food-related content, can offer recipes as well as images of available dishes. These global platforms also allow more insect-experienced countries to encourage the adoption of entomophagy in western countries. The efficacy of this strategy is limited by the notion that sharing of entomophagy may be limited by the fear that it will generate a prejudice towards them [33]. Social media methods create enhanced **observability** by endorsements from established food pages or celebrities. Some do distrust information from these sources and these endorsements may only need to be reflective of the possible lifestyle with limited focus on information distribution. Social media also allows for peer-group pressure to influence spending of certain age groups (such as teenagers) on insect proteins. This will be important as people are more likely to try an insect-containing product when it is offered by a friend than an unknown individual [7, 34, 35].

3.4 Relative advantage

A lack of necessity is a barrier to entomophagy uptake in westernised countries and countries with high meat production and consumption may perceive a lack of need for meat-based alternatives [36]. This highlights a barrier to the introduction strategy of insects as a meat-based alternative as consumers have a food gradient which they follow when selecting meat-based alternatives with initial choice being fish and eventual choices including tofu and similar products at the bottom. Although not a linear path, this gradient shows that consumers have a hierarchy of foods that they follow with novel foods often situated at the bottom [25]. One proposed strategy to overcome this lack of relative advantage would be to avoid promotion as meat alternative. Instead, comparison to nuts could prove more productive as they share similarities in texture, macronutrient content, flavour and size and will circumnavigate the problem that insects encounter when replacing dishes with larger portions of meat such as steaks [32]. In order to fully capitalise on a **relative advantage** over other products, the environmental benefits can be emphasised. Most current comparisons, however, are with meat and there is still debate surrounding this area with vegetarian diets becoming ever more popular in western countries [37, 38]. If the environmental benefit argument is to be made, using circular production gives some insect products **relative advantage** over other 'green' alternatives. A total of 1.3 billion tonnes of food produced for human consumption is wasted per year and valorisation can occur through the use of certain insect species to convert this wasted food into a high-protein product to be used for human or livestock consumption [39, 40].

Differences among western populations affect uptake of entomophagy as individual cultures place different values on factors when choosing their food. For example, the French place a higher value on the pleasures and the social aspects of food consumption whereas the English favour convenience, organic and ethical issues when choosing their food options [41]. With entomophagy, French respondents place less value on the relative advantages of insect products and the British have been found to be more repulsed by visible insects. Understanding this variation creates an opportunity to have adaptive introduction models in different countries. This approach will work to increase the **compatibility** of entomophagy and could also be used to adapt legislation within different legal jurisdictions [42].

4. Availability of product and information

4.1 Absence of available products

The lack of general availability of insect products creates greater complexity through reducing the ease of both **trialability** and enduring adoption. In some cases, demand may already exceed supply and the currently rising visibility will influence social norms causing an increase in demand [32]. As seen with sushi and lobster, greater observability and supply can change societal views and there is no reason this could not be the same for entomophagy [32, 43]. A great range of

Recipe type	Insect	Nut	Chicken	Vegetable	Biscuit
Number of 'hits' (millions)	68	176	991	1960	2550
Relative abundance (to insect)	1	2.6	15	29	38

Table 1.
The number and relative abundance to 'insect recipe' as reference, of 'hits' (search results) in response to the search terms 'insect recipe', 'nut recipe', 'chicken recipe', 'vegetable recipe' and 'biscuit recipe'. Searches conducted using Google Chrome, 21 June 2019.

well-presented products are now available and this very variety can increase acceptance and adoption by consumers [24]. In addition to this, having a wider variety available can reduce the stigma insects have with their strong associations to pests or to notably high-revulsion species such as cockroaches [16].

4.2 Absence of available information

The limited supply of appropriate resources for the sourcing and preparation of insect-containing dishes adds to **complexity**; people do not know where to find recipes, choice advice and cooking information [36]. Though this is now changing rapidly, there remains a substantial information deficit. In 2015, recipes using pine needles and whale meat were more common than insect recipes on the food website 'food.com' and there are currently almost 40 times more mentions of biscuit recipes than insect recipes in a goggle search (**Table 1**) [32].

Along with a knowledge deficit, there is also a confidence deficit contributing to complexity as many people would rather try insects for the first time in a restaurant setting than at home [25]. It is clear that to move towards **lower complexity**, there is a need for an increase in the availability of accessible and free resources. To reduce the need for extensive research, social media, online repositories and increased product information and recipes on packaging all have a part to play. These routes can encourage expansion of the range of dishes individuals will be willing to trial, and, through increasing **trialability** in this way, it can reduce the overall complexity associated with entomophagy.

5. Absence of relative advantage through high prices

Increasing the availability of insect products alone will not be sufficient to drive consumer acceptance for enduring entomophagy. Of those participating in a Dutch study, one-third found insect products to be 'prohibitively expensive' and although most people said price alone would not stop them from purchasing, the remaining two-thirds did recognise price as a factor in repeat purchase decisions [44]. In the 2019 online market place, insect protein powders are 3–10 times the price of vegetable and dairy comparators. Many things currently affect sale price and the increased production now happening across Europe and the North American continent will act to reduce this. Quality, reliability and cost effectiveness arising from increased automation and appropriate species selection will help to reduce price and mitigate the current absence of relative advantage [3].

6. Conclusion

Though interest and product availability are rising, western society has yet to adopt entomophagy as common practice. Entomophagy remains largely

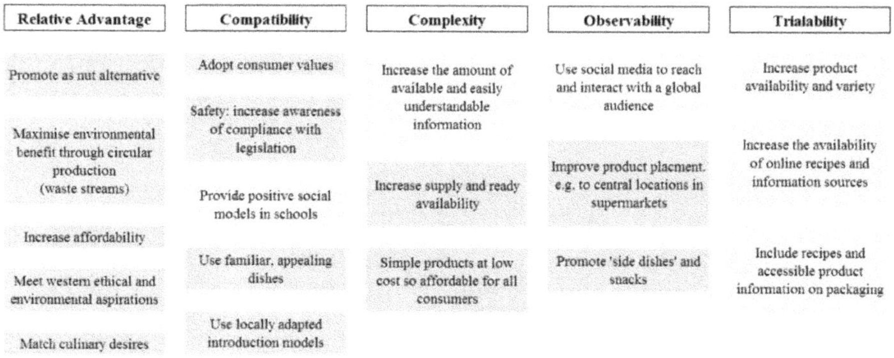

Relative Advantage	Compatibility	Complexity	Observability	Trialability
Promote as nut alternative	Adopt consumer values	Increase the amount of available and easily understandable information	Use social media to reach and interact with a global audience	Increase product availability and variety
Maximise environmental benefit through circular production (waste streams)	Safety: increase awareness of compliance with legislation		Improve product placment. e.g. to central locations in supermarkets	Increase the availability of online recipes and information sources
Increase affordability	Provide positive social models in schools	Increase supply and ready availability		
Meet western ethical and environmental aspirations	Use familiar, appealing dishes	Simple products at low cost so affordable for all consumers	Promote 'side dishes' and snacks	Include recipes and accessible product information on packaging
Match culinary desires	Use locally adapted introduction models			

Figure 2.
The relationships between potential strategies to overcome barriers to entomophagy. Boxes of the same colour indicate strategies that share an overarching theme or whose implementation can improve the ability of other strategies to meet their goal.

incompatible with western ideals, and most westerners exhibit a disgust response when faced with the prospect of eating an insect. A lack of social context and awareness increases the complexity of the innovation and is clearly indicated by consumers experiencing high levels of food neophobia or low awareness of purchase and preparation options.

This chapter has outlined a multitude of promising strategies to overcome such barriers and these strategies need to be developed concurrently (**Figure 2**). When combined, they may help ensure that entomophagy has each of the five characteristics outlined by Rogers as influential to product penetration (**Figure 1**) [8].

Though many recent studies reviewed here have found an increase in participant interest and willingness to adopt through the provision of experience with entomophagy, more research on long-term adoption is required. We need to understand what will embed long-term adoption after food neophobia and the disgust response have been attenuated.

Acknowledgements

The authors wish to thank the many people who contribute to advancing entomophagy and the many benefits it may convey.

Conflict of interest

The authors declare no conflict of interest.

Author details

Harry McDade[1] and C. Matilda Collins[2*]

1 Department of Life Sciences, Imperial College London, London, United Kingdom

2 Centre for Environmental Policy, Imperial College London, London, United Kingdom

*Address all correspondence to: t.collins@imperial.ac.uk

IntechOpen

References

[1] Trends Tracker (Blueshift Research). Interest in Insect-Based Products Grows Slightly but Hovers Around One-Third of Respondents. The Other Two-Thirds are Awaiting Recommendations, Confirmation of Nutritious Value and Taste [Internet]. 2015. Available from: http://blueshiftideas.com/content/june-2015-trends-tracker/

[2] Ahuja K, Deb S. Edible Insects Market Size By Product, By Application, Industry Analysis Report, Regional Outlook, Application Potential, Price Trends, Competitive Market Share & Forecast,. 2018. pp. 2018-2024

[3] van Huis A. Potential of insects as food and feed in assuring food security. Annual Review of Entomology. 2013;**58**(1):563-583

[4] Caparros Megido R, Sablon L, Geuens M, Brostaux Y, Alabi T, Blecker C, et al. Edible insects acceptance by Belgian consumers: Promising attitude for entomophagy development. Journal of Sensory Studies. 2014;**29**(1):14-20

[5] Berenbaum MR. A consuming passion for entomophagy. American Entomologist. 2016;**62**:140-142

[6] van Huis A, Oonincx DGAB. The environmental sustainability of insects as food and feed. A review. Agronomy for Sustainable Development. 2017;(1):37-43

[7] Collins CM, Vaskou P, Kountouris Y. Insect food products in the Western world: Assessing the potential of a new 'green' market. Annals of the Entomological Society of America. 2019. Special Issue: Edible Insects 1-11

[8] Rogers EM. Diffusion of Innovations. 5th ed. New York: Simon & Schuster; 2003

[9] Chapman HA, Anderson AK. Understanding disgust. Annals of the New York Academy of Sciences. 2012;**1251**(1):62-76

[10] Hartmann C, Shi J, Giusto A, Siegrist M. The psychology of eating insects: A cross-cultural comparison between Germany and China. Food Quality and Preference. 2015;**44**:148-156

[11] Rozin P, Fallon AE. A perspective on disgust. Psychological Review. 1987;**94**(1):23-41

[12] Hamerman EJ. Cooking and disgust sensitivity influence preference for attending insect-based food events. Appetite. 2016;**96**:319-326

[13] La Barbera F, Verneau F, Amato M, Grunert K. Understanding Westerners' disgust for the eating of insects: The role of food neophobia and implicit associations. Food Quality and Preference. 2018;**64**:120-125

[14] Tuorila H, Lähteenmäki L, Pohjalainen L, Lotti L. Food neophobia among the Finns and related responses to familiar and unfamiliar foods. Food Quality and Preference. 2001;**12**(1):29-37

[15] Rozin P, Millman L, Nemeroff C. Operation of the laws of sympathetic magic in disgust and other domains. Journal of Personality and Social Psychology. 1986;**50**(4):703-712

[16] Deroy O, Reade B, Spence C. The insectivore's dilemma, and how to take the west out of it. Food Quality and Preference. 2015;**44**:44-55

[17] Siegrist M. Factors influencing public acceptance of innovative food technologies and products. Trends in Food Science & Technology. 2008;**19**(11):603-608

[18] Earle T, Siegrist M. Trust, confidence and cooperation model:

A framework for understanding the relation between trust and risk perception. International Journal of Global Environmental Issues. 2008;**8**(1):17

[19] Siegrist M, Cousin ME, Kastenholz H, Wiek A. Public acceptance of nanotechnology foods and food packaging: The influence of affect and trust. Appetite. 2007;**49**(4):459-466

[20] Belluco S, Losasso C, Maggioletti M, Alonzi CC, Paoletti MG, Ricci A. Edible insects in a food safety and nutritional perspective: A critical review. Comprehensive Reviews in Food Science and Food Safety. 2013;**12**(3):296-313

[21] Finke MD, Rojo S, Roos N, van Huis A, Yen AL. The European food safety authority scientific opinion on a risk profile related to production and consumption of insects as food and feed. Journal of Insects as Food and Feed. 2015;(1):245-247

[22] Gmuer A, Nuessli Guth J, Hartmann C, Siegrist M. Effects of the degree of processing of insect ingredients in snacks on expected emotional experiences and willingness to eat. Food Quality and Preference. 2016;**54**:117-127

[23] Pliner P, Pelchat M, Grabski M. Reduction of neophobia in humans by exposure to novel foods. Appetite. 1993;**20**(2):111-123

[24] Wansink B. Changing eating habits on the home front: Lost lessons from World War II research. Journal of Public Policy & Marketing. 2003;**21**(1):90-99

[25] Schösler H, de Boer J, Boersema JJ. Can we cut out the meat of the dish? Constructing consumer-oriented pathways towards meat substitution. Appetite. 2012;**72**:1592-1607

[26] Peryam DR, Polemis BW, Kamen JM, Eindhoven J, Pilgrim FJ. Food Preferences of Men in the U. S. Armed Forces. Surveys of Progress on Military Subsistance Problems. Chicago: Quartermaster Food and Container Institute; 1960

[27] Martins Y, Pliner P. Human food choices: An examination of the factors underlying acceptance/rejection of novel and familiar animal and nonanimal foods. Appetite. 2005;**45**(3):214-224

[28] Higgs S, Thomas J. Social influences on eating. Current Opinion in Behavioral Sciences. 2016;**9**:1-6

[29] Jensen NH, Lieberoth A. We will eat disgusting foods together—Evidence of the normative basis of Western entomophagy-disgust from an insect tasting. Food Quality and Preference. 2019;**72**:109-115

[30] Wadhera D, Capaldi Phillips ED, Wilkie LM, Boggess MM. Perceived recollection of frequent exposure to foods in childhood is associated with adulthood liking. Appetite. 2015;**89**:22-32

[31] Birch LL, Marlin DW. I don't like it; I never tried it: Effects of exposure on two-year-old children's food preferences. Appetite. 1982;**3**(4):353-360

[32] Shelomi M. Why we still don't eat insects: Assessing entomophagy promotion through a diffusion of innovations framework. Trends in Food Science & Technology. 2015;**45**:311-318

[33] Looy H, Dunkel FV, Wood JR. How then shall we eat? Insect-eating attitudes and sustainable foodways. Agriculture and Human Values. 2014;**31**(1):131-141

[34] Lensvelt EJS, Steenbekkers LPA. Exploring consumer acceptance of entomophagy: A survey and experiment in Australia and the Netherlands.

Ecology of Food and Nutrition. 2014;**53**(5):543-561

[35] Spero I, Stone M. Agents of change: How young consumers are changing the world of marketing. Qualitative Market Research: An International Journal. 2004;**7**(2):153-159

[36] Clarkson C, Mirosa M, Birch J. Consumer acceptance of insects and ideal product attributes. British Food Journal. 2018;**120**(12):2898-2911

[37] Oonincx DGAB, de Boer IJM. Environmental impact of the production of mealworms as a protein source for humans—A life cycle assessment. PLoS One. 2012;7:12

[38] Shelomi M. The meat of affliction: Insects and the future of food as seen in expo 2015. Trends in Food Science and Technology. 2016;**56**:175-179

[39] Gustavsson J, Cederberg C, Sonesson U. Global Food Losses and Food Waste—Extent. Causes and Prevention. Rome, Italy: FAO; 2011

[40] van Huis A, Klunder JVIH, Merten E, Halloran A, Vantomme P. Edible Insects. Future Prospects for Food and Feed Security. Rome: Food and Agriculture Organization of the United Nations; 2013

[41] Pettinger C, Holdsworth M, Gerber M. Psycho-social influences on food choice in southern France and Central England. Appetite. 2004;**42**(3):307-316

[42] Halloran A, Muenke C, Vantomme P, van Huis A. Insects in the human food chain: Global status and opportunities. Food Chain. 2014;**4**(2):103-118

[43] Luzer D. How Lobster Got Fancy. Pacific Standard [Internet]. 2013. Available from: https://psmag.com/

[44] House J. Consumer acceptance of insect-based foods in the Netherlands:

Academic and commercial implications. Appetite. 2016;**107**:47-58

Chapter 3

Larval Development and Molting

Amritpal Singh Kaleka, Navkiran Kaur
and Gaganpreet Kour Bali

Abstract

The term larva applies to the young hatchling which varies from the grown up adult in possessing organs not present in the adult such as sex glands and associated parts. Insect development is of four types namely Ametabolous, Paurometabolous, Hemimetabolous and Holometabolous. The larvae appear in variety of forms and are termed as caterpillars, grubs or maggots in different insects groups. The larval development consists of series of stages in which each stage is separated from the next by a molt. It's a complex process involving hormones, proteins and enzymes. Insects grow in increments. The molting is the process through which insects can routinely cast off their exoskeleton during specific times in their life cycle. The insect form in between two subsequent molts is termed as instar. The number of instars varies from 3 to 40 in different insect orders depending on the surrounding environmental and other conditions such as inheritance, sex, food quality and quantity. The larvae are categorized into four types namely Protopod larva, Polypod larva, Oligopod larva and Apodous larva.

Keywords: instar, insect, larval development, molting

1. Introduction

A larva is a distinct immature developmental form of many animals particularly in insects. The term larva applies to the young hatchling which varies from the grown up adult in lacking some important organs like sex glands and associated parts. The animals such as insects, amphibians and cnidarians with indirect development typically have larval phase in their life cycle. The diet of the larva is considerably distinct from the adult.

The larval forms are often adapted to different environments than of adults. For example, larvae of mosquitoes live almost exclusively in aquatic environment during their developmental stages and live outside water after metamorphosing into adult forms. Such adaptations in distinct environments are for their protection from predators and to avoid competition for resources. During developmental stages, larvae consume more food to fuel up their transition into adult form. In some insect species immature forms are totally dependent on adult forms for feeding such as in social insects of orders Hymenoptera (e.g., bees, wasps, ants) and Isoptera (e.g., termites) and the female workers feed them.

There are several advantages of an embryo developing into larval form instead of growing into an adult directly because it would help the animal to overcome various difficulties. The hatchling may need to obtain food but due to its smaller size is unable to feed itself the same way as an adult does. Further, it would be unable to make an effective use of defense mechanism as done by adult. Thus, the new organization of freshly emerged organism is best suitable to its environment. Further it furnishes a mode of life which is better suited to newly emerged small hatchling. The additional advantage of this corresponding organization is that it enables the larva to exploit an entirely different environment from that of its grown-ups. Thus, a terrestrial adult may have aquatic larval form such as in order Odonata (Dragonflies and Damselflies), a flying adult may have burrowing larvae as in order Diptera (Flies) and an adult may have free-living larvae in order Trichoptera (Caddisflies).

The arthropods cast off their cuticle at regular intervals to undergo a brief period of development before reaching mature size. Post-embryonic development is divided into a series of stages in which each stage is distinct from the next one by a molt. The larval forms usually change in shape during their development and progressive stages are not similar in insects. This change in form is known as metamorphosis. These changes are controlled by a juvenile hormone which is secreted by glands-corpora allata present in the posterior region of insect's head. It is released during each molt and its amount decreases each time. As the concentration of juvenile hormone declines, more adult characters appear and the adult stage is produced. In arthropods, the larval forms move between stages by molting of their exoskeleton. The new exoskeleton develops beneath the old skin. During the formation of new exoskeleton, insect's body gets swelled up due to intake of either air or water until the old exoskeleton breaks down. The newly formed exoskeleton hardens and different tanning agents get deposited onto the surface. After the succession of molts, an insect reaches the final adult form and no further molt takes place. Each developmental stage of an arthropod between molts is termed as Instar. For example, after hatching from egg, the hatchling is said to be first instar. When the insect molts again, it is then a second instar and so on.

2. Patterns of growth and development in insects

There are four patterns of growth and development in insects namely Ametabolous, Paurometabolous, Hemimetabolous and Holometabolous.

2.1 Ametabolous development (simple metamorphosis)

It is the type of insect development in which there is no metamorphosis. The emerged immature stage appears very similar to adult except that it lacks sexual structures. It grows only in size by replacing its old skin through molting. The larva grows bigger and the genitalia develops progressively with each molt. The young one which emerges from egg resembles adult in miniature form, is called nymph. The reproductive organs are undeveloped in nymph and after molting the nymph becomes an adult. Both forms i.e., the nymphs and adults live in the same habitat.

This is the characteristic feature of Apterygotes (e.g., Silverfish-*Lepisma* Linnaeus and Springtail). For instance, the silverfish hatched from egg looks like an adult and undergoes subtle anatomical changes between molts (**Figure 1**). Immature silverfish molts 6–7 times until it reaches sexually mature adult stage. In favorable conditions, silverfish may typically continue to molt during its lifespan and molts 25–66 times [1].

2.2 Paurometabolous development (gradual metamorphosis)

Paurometabolous development is found in less primitive forms like cockroaches, grasshoppers, praying mantis and white ants. In this type of development, the

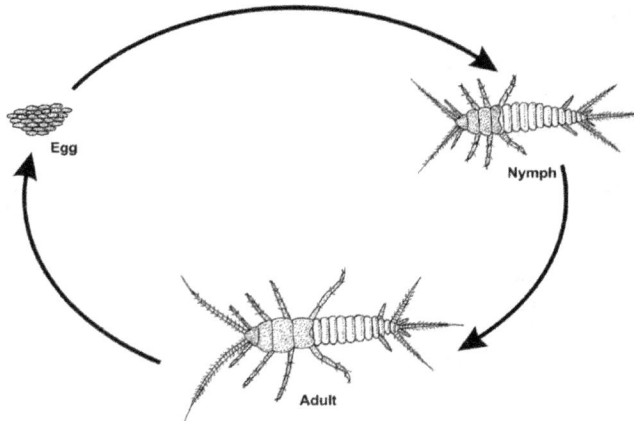

Figure 1.
Ametabolous development in Lepisma.

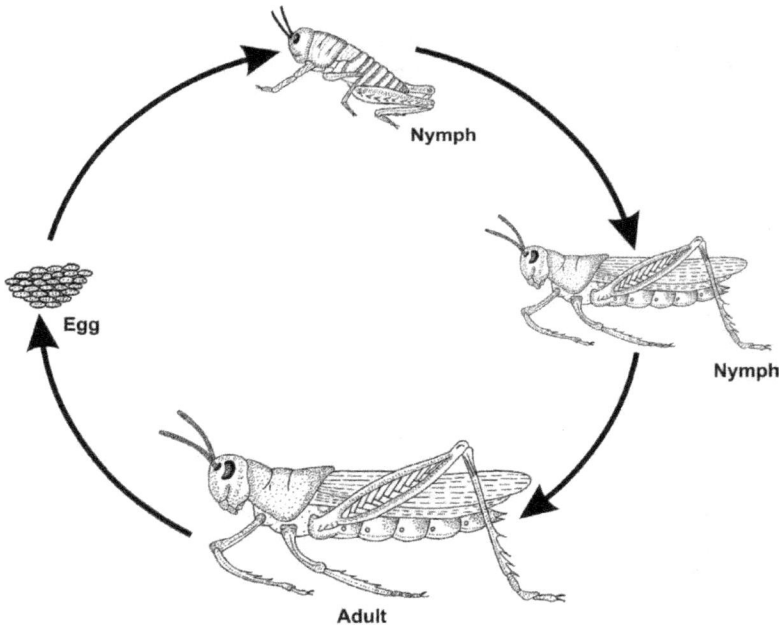

Figure 2.
Paurometabolous development in grass-hopper.

newly emerged young one closely resembles the adult in general body form, habits and habitat but many adult characters like wings and reproductive organs are not developed and their relative proportions of the body also differs. The young forms are termed as nymphs. The wings develop as wing pads on second and third thoracic segments at an early stage and gradually increase in size during each successive molt. The external genitalia also develops gradually after each molt. These nymphs lead an independent life and attain adult features through several molts. There are three stages in the life cycle of these insects i.e., egg, nymph, imago (adult) and no pupal stage is there. For instance in grasshoppers, before becoming adults the nymphs undergo 5–6 molts to change their body form (**Figure 2**). The nymph stage is species specific and lasts for a period of 5–10 days depending upon the weather conditions like temperature and humidity.

2.3 Hemimetabolous development (incomplete metamorphosis)

In this type of development, adult form is attained by gradual morphological changes with successive molts. The hatched larva lacks wings and genitalia but have some other characteristic features which are absent in adult. These features are lost at the final molt. The orders Plecoptera, Ephemeroptera and Odonata (**Figure 3**) have aquatic larval stages. The young forms are known as naiads which are aquatic and respire by external gills but the adults are terrestrial in behavior. Their life cycle also involves three stages: eggs, naiads and adults. When the naiads are ready to transform into adults, they come out of water and adult winged forms are released. The wings and genitalia develop externally but are not fully formed till adulthood. After the formation of wings no further molting takes place, only exception in mayflies where winged forms of aquatic nymphs come out and rest on trees to undergo final molting to become adults.

2.4 Holometabolous development (complete metamorphosis)

Complete metamorphosis is a kind of morphological change during post-embryonic transformation in which larva has no similarity with adult and there is always

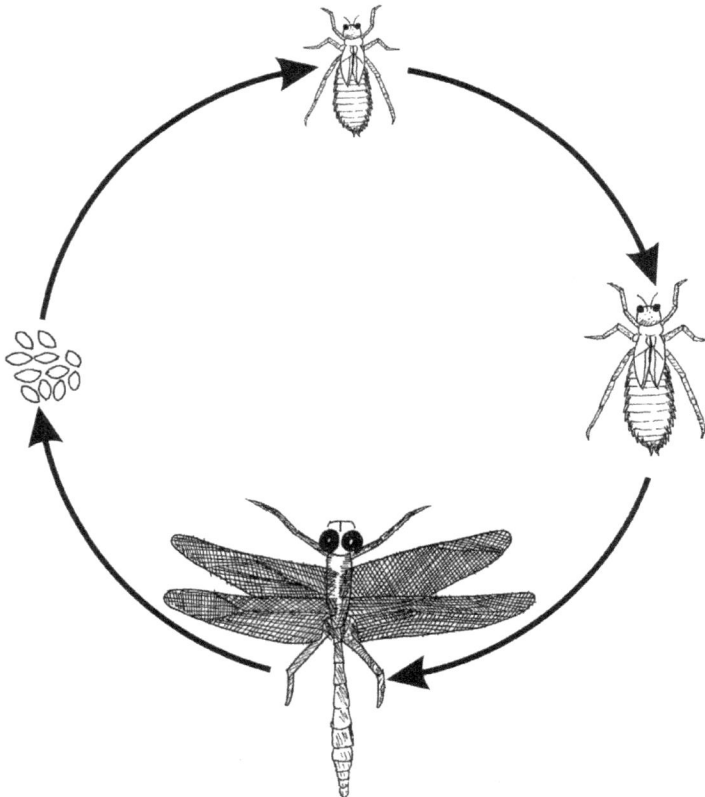

Figure 3.
Hemimetabolous development in dragonfly.

a pupal stage. Complete metamorphosis takes place in orders Coleoptera, Diptera, Hymenoptera and Lepidoptera. Pupal stage is the characteristic of holometabolous development i.e., this stage is present between the last larval stage and the adult.

In Order Lepidoptera (moths and butterflies), the larva is known as Caterpillar (**Figure 4**). It possesses a distinct head with powerful mandibles and three pairs of jointed thoracic legs. The abdomen has four or five pairs of un-jointed, short abdominal legs which are termed as pseudo-legs or prolegs. These caterpillars eat voraciously and grow rapidly with several moltings. After completing four or five molts, the caterpillar is transformed into pupal stage.

In Order Diptera (Houseflies and other flies), the larva is worm-like and devoid of appendages and is known as maggot (**Figure 5**). The mature larva is about 12 mm long. The head is indistinct, with a pair of oral lobes and hooks.

In Order Coleoptera (ground beetles, ladybirds and rove beetles), like adults the larvae referable to many beetle families are predatory in nature. The larval morphology is highly varied among species, with well-developed and sclerotized heads, distinguishable thoracic and abdominal segments and are known as grubs.

In Order Hymenoptera (bees and wasps), the larvae are grub-like with well developed head and mouthparts are of chewing type. Larvae are generally apodous, rarely eruciform with locomotory appendages.

2.5 Types of larvae

The larvae in different orders of insects are known by different names i.e., larvae of butterflies and moths are termed as caterpillars and those of Diptera and

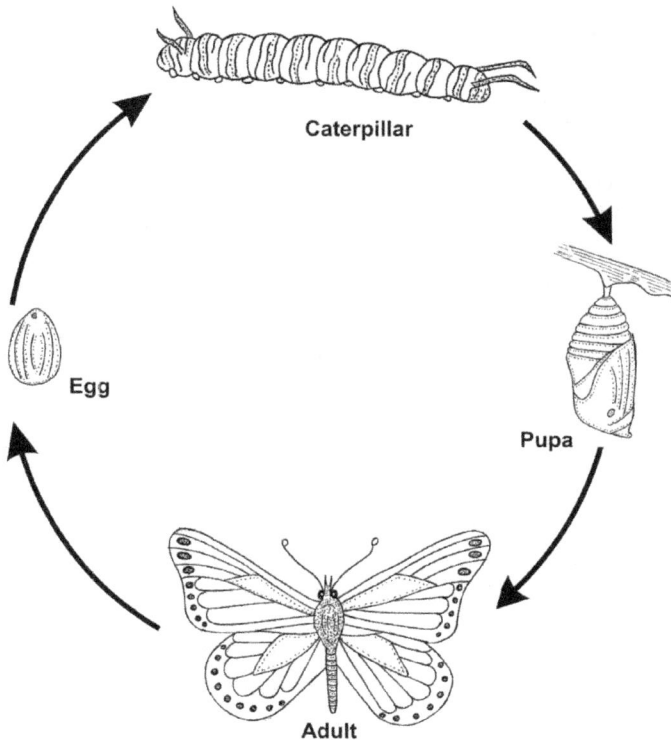

Figure 4.
Holometabolous development in butterfly.

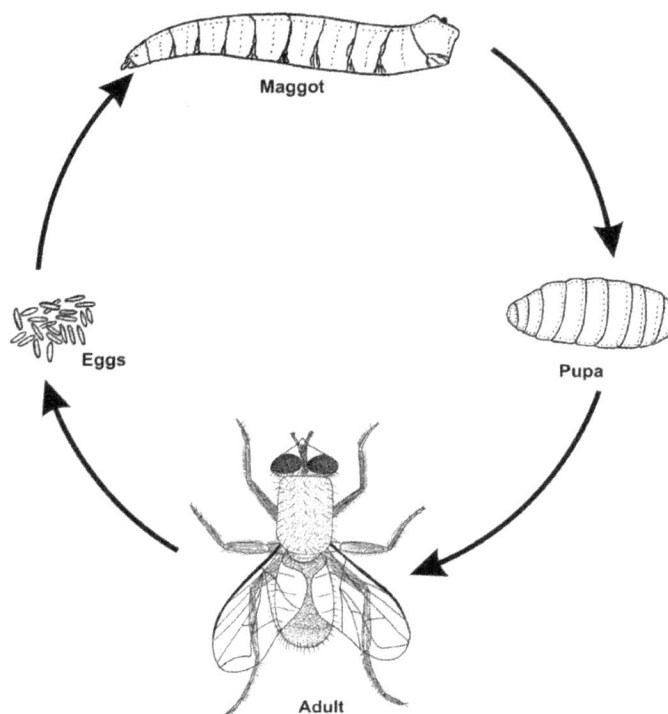

Figure 5.
Holometabolous development in housefly.

Coleoptera are termed as maggots and grubs respectively. The larvae are grouped into four types on the basis of development of appendages (**Figure 6**).

1. **Protopod larva**: In this type, larvae come out from the eggs which contain very little amount of yolk and this happens during the early stages of embryonic development. There is no segmentation on the abdomen. The thoracic appendages and head (cephalic) are primitive in form e.g., endoparasitic larvae of order Hymenoptera.

2. **Polypod larva**: In this type, larvae have three pairs of thoracic legs and two to five pairs of abdominal prolegs. The body of the polypod larva is well segmented and is termed as "eruciform" (cylindrical type). Only prothoracic and abdominal spiracles are open in their respiratory system. Larvae of orders Mecoptera (Scorpion flies and hanging flies) and Lepidoptera (butterflies and moths) are of polypod type.

   On the basis of number and location of prologs, the lepidopteran larvae are further classified into three types: caterpillar, semilooper and looper.

   a. Caterpillar: Caterpillar is the larval stage in Order Lepidoptera. It has soft body that can grow rapidly between molts. It bears five pairs of prolegs which are present on 3rd, 4th, 5th, 6th and 10th abdominal segments and three pairs of thoracic legs. e.g., larvae of gram pod borer and Lemon butterfly.

   b. Semilooper: The semilooper larva bears three pairs of thoracic legs and three pairs of prolegs which are present on 5th, 6th, and 10th abdominal segments e.g., Cotton Semilooper and Castor Semilooper.

   c. Looper: The looper larva have three pairs of thoracic legs and two pairs of prolegs present on 6th and 10th abdominal segments e.g., Cabbage looper.

3. **Oligopod larva**: The body of the oligopod larva is well segmented. It have three pairs of thoracic legs and possesses well developed cephalic appendages. The prolegs are absent. In some oligopod larvae, a pair of cerci or similar caudal processes is present. Head capsule is well developed and mouthparts are similar to the adult. On the basis of structure, the oligopod larvae can be further classified into two types viz., Campodeiform type and Scarabaeiform type.

   a. Campodeiform type: The campodeiform larva has dorso-ventrally flattened and well sclerotized body which bears long thoracic legs and a pair of terminal cerci. This type of larvae is found in orders Neuroptera, Trichoptera, Strepsiptera and in some Coleoptera (e.g., Lacewing and Ladybird beetle).

   b. Scarabaeiform type: The larva in this type is fleshy and 'C'-shaped with poorly sclerotized abdomen and thorax. It bears short legs and terminal abdominal processes (cerci) are absent. These larvae are less active and sluggish in nature. Scarabaeiform larvae are mainly in Scarabaeoidea and also in some other Coleopterans (e.g., White grub, Rhinoceros beetle).

4. **Apodous Larva**: Thoracic legs or abdominal prolegs are absent in case of apodous larva and it has poorly sclerotized cuticle (e.g., Honey bee, House fly, Fruit fly). On the basis of degree of development of head, the apodous larvae can be further grouped into the following three types:

   a. Eucephalous Larva: In this type, larva has well sclerotized head capsule with relatively reduced cephalic appendages and is found in Nematocera (Diptera), Cerambycidae (Coleoptera) and Aculeata (Hymenoptera). E.g., Mango stem borer, Mosquito.

   b. Hemicephalous Larva: In this type, larva has reduced head capsule that can be withdrawn within the thorax. It is found in families Tipulidae and Tabanidae of order Diptera. E.g., Crane fly, Horse fly.

   c. Acephalous Larva: This type of larva has no head capsule and cephalic appendages. E.g., Larva of House fly).

Like larvae, the pupae are also of various types (**Figure** 7). These can be grouped according to the presence or absence of functional mandibles which might be used by the adult to emerge from the cocoon or pupal cell. The functional mandibles are present in decticous type of pupa, whereas in the adecticous type, the mandibles are not functional. The latter type can be subdivided into two: exarate and obtect. The exarate pupa has free appendages and the obtect have appendages glued to the rest of the pupal body. An exarate pupa enclosed in a puparium is termed as coarctate whereas the silken protective case of obtect pupa is known as cocoon.

2.6 Heteromorphosis

Heteromorphosis is the type of development characterized by radical change in forms between successive larval instars. The larval instars are pretty much similar in many endopterygotes. However, a larva experiences typical change in morphology

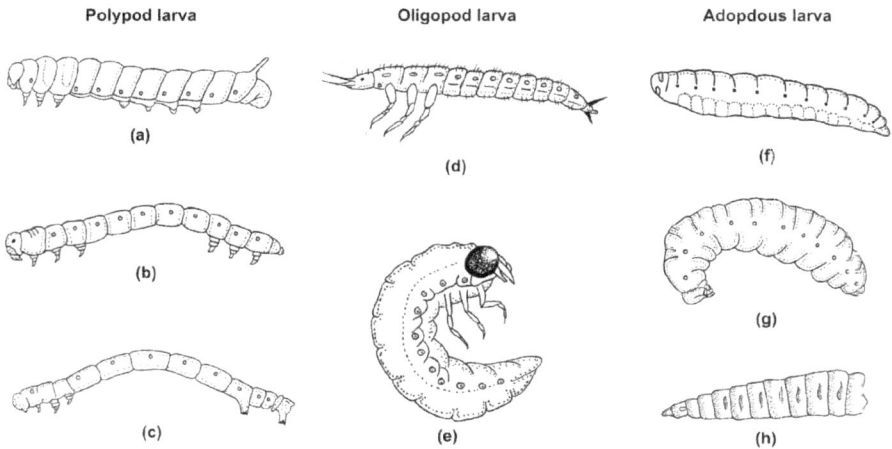

Figure 6.
Different types of insect larvae: (a) Caterpillar, (b) Semilooper, (c) Looper, (d) Campodeiform, (e) Scarabaeiform, (f) Eucephalous larva, (g) Hemicephalous larva, (h) Acephalous larva.

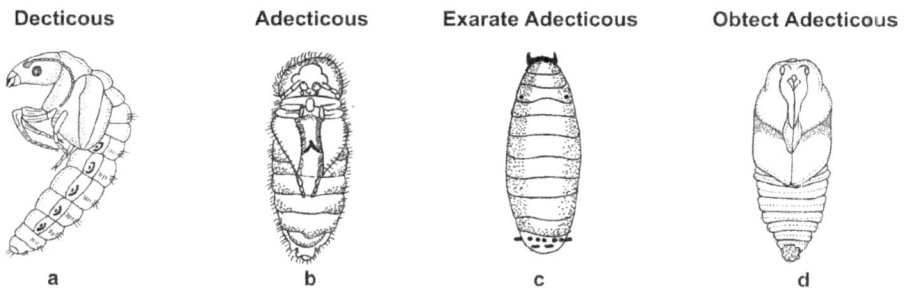

Figure 7.
Different types of insect pupae: (a) Decticous pupa (b) Adecticious pupa, (c) Exarate Adecticous pupa, (d) Obtect Adecticous pupa.

and in habits during development in some families of orders Coleoptera (Meloidae, Ripiphoridae), Diptera, Hymenoptera, Neuroptera and in all Strepsiptera. It is common in parasitic and predaceous insects where change in habit occurs during course of development. This is further of two types:

In first type, eggs are laid in open and the first stage larva searches for its host. In this type, the newly emerged first stage larva is an active campodeiform larva. For example, in Strepsiptera, larva attaches itself to a host often a bee or a sucking bug. When the bee visits a flower in which the larva is lurking, subsequently, it becomes an internal parasite and loses all traces of appendages and a series of dorsal projections starts developing which increases its absorptive area. The cephalothorax develops during sixth and seventh larval stages. Another example is of blister beetles (Meloidae), the larva hatches as free-living campodeiform which can actively search for food. After locating the food source, the larva soon molts to second stage i.e., eruciform (caterpillar like). Further, it has to pass through either two or more additional larval instars, where it may remain as eruciform or become scarabaeiform. A basically similar life history with an active first stage larva followed by inactive parasitic stages occur in Acroceridae (Diptera), Bombyliidae, Epipyropidae (Lepidoptera), Mantispidae (Neuroptera), Nemestrinidae (Diptera), Perilampidae, Eucharidae (Hymenoptera), Meloidae and some Staphylinidae (Coleoptera).

In second type of heteromorphosis, the eggs are laid in or on the host. It occurs in some endoparasitic Diptera and Hymenoptera. Among the parasitic Hymenoptera, the newly emerged larva is of protopod type larva. It has many different forms in different insect species. For instance, the first stage larva of *Helorimorpha* Schmiedeknecht of family Braconidae (Hymenoptera) has a small unsegmented body, a big head and a tapering tail. On the other hand the third stage larva is a fairly typical eucephalous hymenopteran larva. In Platygastridae, the first stage larva is more specialized, with an anterior cephalothorax bearing rudimentary appendages. These larvae hatch from eggs which contain very little amount of yolk.

2.7 Number of instars

During immature development, larvae of insects and other arthropods molt regularly by shedding their exoskeletons. Thus, instar is a developmental stage between two successive molts in the life cycle of an arthropod and the development period of larvae in insects is divided into a few discrete stages. The dissimilarity between instars is often observed in altered body proportions, patterns, colors, number of body segments, head width and appendages. Arthropods shed their exoskeleton in order to grow and to assume a new form. After shedding their exoskeleton, the juvenile insects continue their life cycle till they either pupate or molt again. The first larval instar stage begins at hatching and it ends at the first larval molt. In holometabolous insects, the last instar is a phase from final molt to either prepupal or pupal stage or the eclosion of an imago in hemimetabolous insects. The period of growth is species specific and is fixed for every instar. The larval instars number varies across various insect species.

An insect instar number also depends on the surrounding environmental and other conditions. The most common factors affecting the number of instars are temperature, humidity, photoperiod, physical condition, inheritance, sex, food quality and quantity. Lower temperature and less humidity often slow down the rate of development. In some insects, e.g., in salvinia stem borer moth, the number of instars relies on early larval nutrition. In addition, the presence of injuries has been observed to influence the number of instars in some species. During suitable conditions, the instar number is higher in exceptional species of orders Orthoptera and Coleoptera. Intraspecific variability in the number of larval instars is a widespread phenomenon occurring in most major insect orders, in both hemimetabolous and holometabolous insects. For instance, the hymenopteran egg parasitoid *Trichogramma australicum* (Girault) have only one larval instar [2], whereas 34 larval instars are reported in *Leptophlebia cupida* (Say) [3]. In some phylogenetically older orders like Plecoptera, Ephemeroptera and Odonata larvae have heavily sclerotized, non-expansible exoskeletons and the instar number is usually 10 [4, 5]. The larvae of a notodontid moth have an additional instar even when exposed to artificial rainfall [6].

Apart from other environmental factors, the inheritance and sex are the two factors which most commonly influence the instar number. The number of instars is usually sexually dimorphic and the females in general have a higher number of instars than males. The inherited factors affecting number of instars may be either hereditary or achieved by means of maternal impacts and further may rely upon environmental conditions encountered by a parent. Instar number might be genetically unique in larvae from various populations [7], between genetically determined phenotypes or between the offsprings of different individuals from the same population. Moreover, instar number may also differ genetically between subspecies [8], or between short and long winged individuals [9]. The instar number of progeny is influenced by the prevailing ecological conditions during the oviposition

Orders	Number of larval stages
Siphonaptera	3
Phthiraptera	3–4
Coleoptera	3–5
Hemiptera	3–5
Neuroptera	3–5
Diptera	3–6
Hymenoptera	3–6
Mecoptera	4
Zoraptera	4–5
Dermaptera	4–6
Embioptera	4–7
Lepidoptera	5–6
Thysanoptera	5–6
Trichoptera	5–7
Mantodea	5–9
Isoptera	5–11
Orthoptera	5–11
Psocoptera	6
Blattodea	6–10
Grylloblattodea	8
Phasmida	8–12
Thysanura	9–14
Ephemeroptera	20–40
Plecoptera	22–23

Table 1.
Number of larval stages in different orders of insects.

or larval period of parents. When reared in isolation, the nymphs of locusts namely *Schistocerca gregaria* Forsskal and *Nomadacris septemfasciata* (Serville) have more instars. In case of gypsy moths, the larvae which develop from the last laid smaller eggs of specific females usually have more instars [10]. There are some further factors which may particularly affect the number of instars. In case of termites, the larvae belonging to different castes also have different number of larval instars [11]. Lycaenid butterfly larvae have more instars when they live in association with ants and they have low pupal weight when ants are not present [12].

The larvae enter in a stage of inactivity i.e., remain motionless after final instar and stop feeding. This stage is known as pupal stage (chrysalis in case of butterflies). The larvae begin to resemble adults at the end of this stage due to the anatomical modifications that take place in them and also due to the appearance of new organs and tissues (**Table 1**).

3. Molting during post-embryonic development

In larval forms, when the exoskeleton is outgrown, the insects undergo molting regularly. In insects, the unique process of molting is under hormonal control and thus involves various hormones, proteins and enzymes.

During developmental phase when an immature insect needs a larger exoskeleton, the neurosecretary cells present in the brain are activated. It helps the larva to ward off its old exoskeleton. Thus molting is the phenomenon by which insects develops. Under controlled and protected conditions, it permits the body of the developing insect to expand. In order to increase in size the insect must sheds its skin in favor of new underneath skin. Insect can molt 5–60 times in the total life span depending upon its species. Stadium is the time interval between the two subsequent molts and instar is the form assumed by the insect in any stadium. Each instar ends with molting.

When there is no more space for the insect to expand inside its old exoskeleton, hormone triggers molting which separates the exoskeleton from the underlying epidermis and the molting fluid fills the newly created gap. Proteins are secreted by the epidermal cells to form a new cuticle. Later on, when the new cuticle is in place, the inner layer of the exoskeleton is digested by the enzymes present in the molting fluid. Epidermal cells recycle the chitin and proteins which are then secreted under the new cuticle.

With the formation of new exoskeleton the insects can further start shedding its old exoskeleton. The insect expands its body with the intake of large amount of air and the outer shell is forced to get split, usually down the dorsal side as a result of muscular contractions. The outgrown exoskeleton squeezes out the bud. It is a compulsion for the insect to expand and swell its newly formed cuticle which conclusively makes this new cuticle large enough so it can allow room for any further growth. The newly formed overcoat is much paler in appearance and is soft than that of the older one, however, it starts to become darker and hardens itself within few hours. The appearance of the insect seems like a slightly larger copy of its previous form.

The whole procedure of development of an insect is influenced mainly by three hormones: Prothoracicotropic hormone (PTTH), Ecdysone and Juvenile hormone which are secreted by neuro-secretory cells (NSC) present in the brain, Prothoracic gland (JH) and corpora allata respectively. The signals are sent by the developing body of the insect to the brain and direct it to activate the clusters of neurosecretory cells which then produce PTTH which passes down into neurohemal organ, Corpora Cardiaca (CC) to release stored PTTH into the circulatory system (**Figure 8**). The prothoracic glands get stimulated by this to secrete Ecdysone. The active form of ecdysone triggers a series of physiological events leading to the formation of a new exoskeleton by the process known as apolysis.

Along with serving the purpose of acting as a hormone release site, the Neurohemal organ also synthesizes hormones. It is the responsibility of JH to maintain the insect in its young state and it modifies expression of the molt, acts in conjunction with ecdysone. JH hormone favors the synthesis of larval structures and adult differentiation and thus considered as a modifying agent.

3.1 Regeneration

During larval development some insects are able to regenerate their appendages following their accidental loss. If the loss occurs early in a developmental stage, before the production of molting hormone, the appendage reforms at the next molt. The regeneration occurs in larval forms of Blattodea, Phasmatodea, in some Hemiptera, Orthoptera and holometabolous insects. The regeneration of cuticular structures can only happen at a molt as this is the only time at which new cuticle is produced. Consequently, regeneration of appendages does not occur in adults and is restricted to larval stages only. Regeneration of muscles and parts of the nervous system also occurs during development stages.

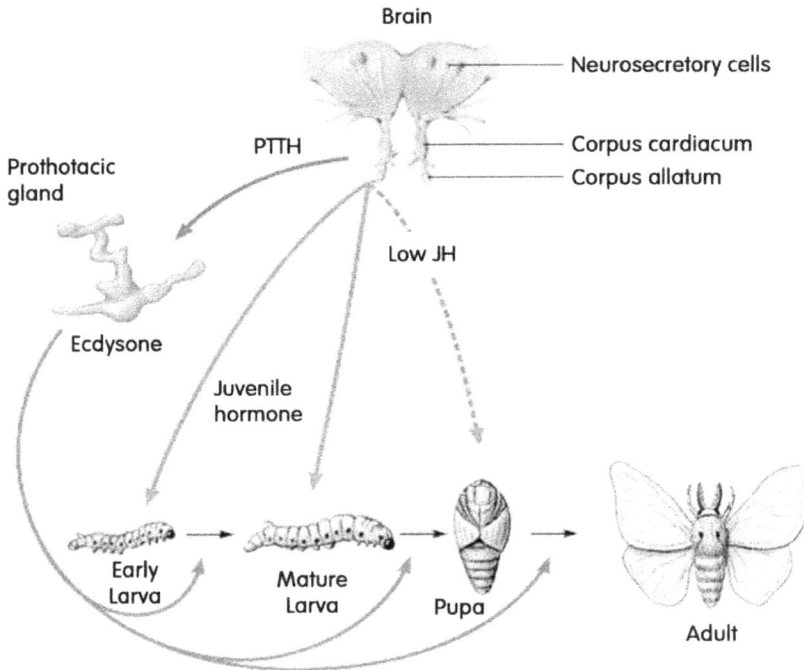

Figure 8.
Hormonal control of insect development.

4. Growth

In an organism, body size is one of the most important life history characters. Its effects on fitness are well documented and have been extensively studied both theoretically and empirically. Gaining weight and eating is the foremost function of the larva which leads to several developmental changes during this phase. The eyes, palpi, proboscis, antennae, reproductive organs and wings starts developing and the larval legs will develop into the adult legs. Prior to pupation, growth of these organs accelerates during the last one or 2 days, hence when pupa is formed the major important changes to the adult form have already been taken place. There is a progressive increase in weight throughout the developmental stages. Body size is flexible i.e., it can change in response to different environmental conditions. For example, insects developing at higher temperatures are generally smaller than those developing at lower temperatures and well fed organisms are typically larger than those fed at poor quality diet.

Normally the weight increases consistently throughout the phase of development and then falls slightly at the time of molting due to the loss of cuticle and some loss of water that is not replaced because the insect is not feeding. After molting, the weight quickly increases above its previous level. Expressed in terms of increase in absolute weight, the growth rate is usually greater in the later stages, but the relative growth rate normally decreases as the organism increases in size. In some aquatic insects before a molt there is no decrease in weight, but, at the same time, there is a sharp increase at the time of ecdysis due to the retention of water, either through the cuticle or by means of alimentary canal. This is utilized to build the volume of the insect, thus splitting the old cuticle.

D'Amico *et al.* analyzed the physiological basis of body size evolution in the tobacco hornworm, *Manduca sexta* (Linnaeus) of family Sphingidae. In this species, the final body size of the adult is determined by five variable factors: the initial size of the last larval instar, growth rate during that instar, critical weight, time delay between achieving critical weight and initiation of prothoracicotropic hormone (PTTH) secretion and timing of photoperiodic rate for PTTH secretion. They demonstrated that in a continuous laboratory culture over a 30 year period (approximately 220 generations) body size of this insect increased by 50%. This evolutionary increase in body size could be accounted for by changes in three of the five factors: growth rate, critical weight, and PTTH delay time (the remaining two factors remained unchanged during this period). The quantitative change in these three factors was shown to account for over 95% of the evolutionary change in body size [13].

In final (fifth) instar larvae of *Manduca sexta* (Linnaeus), somatic growth is causally associated with the timing of number of endocrine events that induce the onset of pupation and metamorphosis [14]. The growth stops and metamorphosis begins, with the secretion of PTTH and ecdysteroids in the final instar. PTTH and ecdysteroid secretions are inhibited by the presence of juvenile hormone (JH) [15]. The level of JH circulating in the hemolymph is high during the first few days of the instar but drops dramatically when the larva attains a specific critical weight. The larval growth stops when the sequence of endocrine and physiological events initiated by the critical weight culminates in the secretion of ecdysteroids [16].

Holometabolous insects acquire their adult biomass during larval growth. In this manner, food consumption is intense and the fat body enlarges amid larval development. However, they do not feed during metamorphosis and simply exploit the nutrients stored during their larval development. During metamorphosis, the fat body is reconstructed through cellular turnover to the degree that when the adult insect emerges, the fat body has been remolded or is completely replaced [17].

The crowding affects the rate of development and also influences the adult size. Insects from crowded conditions are generally smaller than the others developed in isolation. For example in *Aedes aegypti* (Linnaeus), crowding in the habitat generated lighter pupae, longer larval period and increased mortality [18].

4.1 Control of growth

Larval growth is characterized by periodic molts and to some extent the internal changes are correlated with the molting cycle. Larval growth is regulated by ecdysteroid molting hormone which helps in producing larval characters. While hormones exert an overall controlling influence, local factors are also responsible for controlling the form of particular areas in the larval body. For example, epidermal cells often show distinct polarity secreting cuticle in a form giving an obvious anterior–posterior pattern. In the first stage larva of *Schistocerca* Stal, the cuticular plates associated with each epidermal cell on the sides of abdominal sternites are produced into backwardly pointing spines. Similarly in seed bugs, *Oncopeltus* Stal, a row of spines marks the posterior end of the area of cuticle. The scales also grow out with a particular orientation in butterflies and moths. The experimental manipulation shows that the polarity of the cells within a body segment is produced by a gradient of a diffusible substance known as a morphogen.

In addition to having a specific orientation, cuticular structures are dispersed in regular patterns. For example, in assassin bugs (*Rhodnius* Stal), the larvae bear a number of evenly spaced sensilla. Sensilla are the smallest functional units of insect sensory system and form an essential interface between external and internal sensory environments of the insect. At each molt, these increase in number, new sensilla being formed in the biggest gaps between the existing sensilla. This is consistent with the hypothesis that if the sensilla become widely spaced due to the growth of epidermis, the morphogen accumulates between them. If the concentration of morphogen exceeds a certain threshold, the development of a new sensillum is initiated. The development of sensilla on the cuticle in adults of *Oncopeltus* Stal can be accounted for in a similar way. Where two or more integumental features are present in an integrated pattern, they may be controlled by the same substance. For instance, in *Rhodnius* Stal, it is suggested that a differentiating substance- morphogen in high concentration produces the sensilla and the same substance in low concentration initiates the development of dermal glands, which are thus arranged round each sensillum. In *Drosophila* Fallen, and almost certainly in other insects, the boundaries of the para segments are source of signals that organize the patterning and orientation of associated cellular fields [19].

There is relatively little information on control of growth of the integral organs, but some show cyclical activity which coincides with the molt. In *Rhodnius* Stal, the fat body cells exhibit a marked increase in RNA concentration and mitochondrial number just before a molt and the ventral abdominal intersegmental muscles become fully developed only at this time. In some insect orders, Malpighian tubules increase in numbers; mitosis and development of new tubules are phased with respect to each molt.

4.2 Increase in size of the cuticle

Integument is the external layer of tissue that covers the outer surface of insects and the surfaces of the foregut and hindgut. It is composed of epidermis which is a continuous single layered epithelium, an underlying thin basal lamina and the extracellular cuticle that lies on top of the epidermis. An extracellular layer i.e., the cuticle covers the outer surface of the insect's body. It serves dual function. Firstly, it acts as a skeleton for muscle attachment and secondly, as a protective barrier between the insect and its respective environment.

Depending upon insect species, its developmental stage along with the body region, the cuticular layer may vary in thickness which can range from few micrometers to millimeters. It can be as thin as 1 μm in the hindgut and over gills (Ephemeropteran larvae) and as thick as 200 + μm (elytra of large beetles). The cuticle is highly diverse in their mechanical properties and can be divided into two groups: stiff, hard cuticles and soft, pliant (easily bent) cuticles. It also differs in color and in surface sculpturing, but electron microscopy shows that all types of cuticles are built according to a common plan. It is the essential part of the integument which further has the cuticle producing epidermal cells, sense organs and various glands. Epicuticle, procuticle and subcuticle are three distinct layers of the cuticle (**Figure 9**). The epicuticle being the outermost covers the complete cuticular surface. Procuticle comprises the principle part of the cuticle. Subcuticle is present in between the procuticle and the epidermal cells. It is a thin histo-chemically well-defined layer. This layer serves as the accumulation zone where the newly formed cuticular material is assembled and added to the cuticle that already exists and also binds the cuticle and epidermis. In the early stages of the development of the cuticle, before the insect sheds the cuticle of its previous stage, the amount of protein per unit chitin is much less than in the mature developed cuticle [20].

The single celled layer of epidermis constructs the cuticle. The epidermis effectively stores the lamellate endocuticle in those regions where the cuticle is extensible during the intermolt period. At the apical surface of epidermal cells, the plaques of chitin and protein are discharged at the tips of microvilli. Above the plaques in the extracellular space, the cuticle arises by self-assemblage of chitin microfibrils and

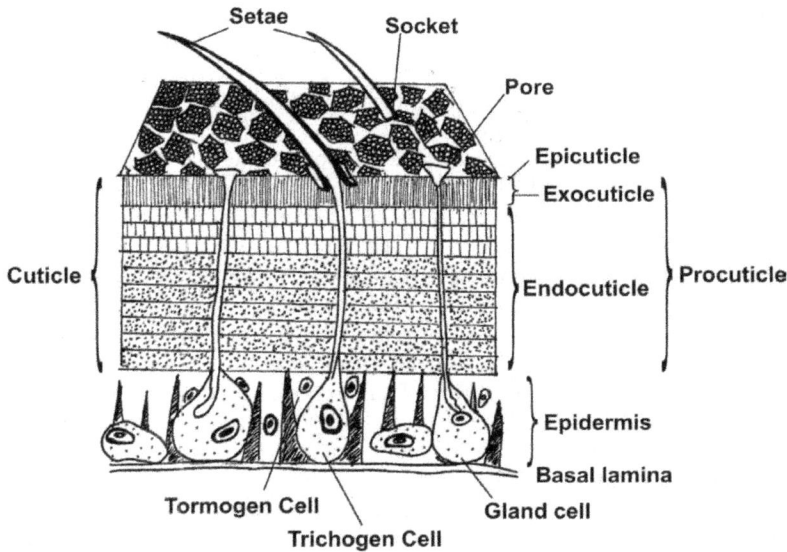

Figure 9.
Structure of insect integument.

secreted proteins. As the larva develops, the epidermal cells beneath the flexible cuticle also grow. The epicuticle in the case of soft-bodied insects e.g., in tobacco hornworm, *Manduca sexta* (Linnaeus) gets deposited in folds to permit development during the succeeding intermolt period. In this case, after ecdysis during the first day, the underlying endocuticle grows by means of apical expansion points created by the discharge of vertical chitin microfilaments. The epidermal cells segregate from the overlying cuticle and experience a burst of RNA synthesis at the beginning of the molt. Endocuticle synthesis comes to an end and then there is secretion of two inactive enzymes namely, chitinolytic and proteolytic. These enzymes form "molting gel" which fills the space in between the apical border of the epidermis and old cuticle. With the activation of these enzymes, the process of digestion takes place in old endocuticle. Then, cuticulin is deposited first at the tips of the plasma membrane plaques followed by deposition between the plaques to form a complete layer. Cuticulin is a structural protein in the cortical layer. It forms the outer part of the epicuticle and provides hardness and waterproofing to cuticle. The epicuticle precursors like polyphenols, lipids and proteins are secreted and congregated on the inner face of the cuticulin layer. Then, the action of phenoloxidases that cross-link the proteins and polyphenols stabilizes the entire structure, followed by the withdrawal of the apical membrane of the epidermal cells from the patterned surface which further starts the formation of procuticle.

4.3 Ecdysis of the old cuticle

Ecdysis is derived from the Greek word *ekdusis*, which means "put off." It describes the process by which arthropods and insects shed/cast off their outer cuticle (exoskeleton). While the new cuticle is formed, the older one remains intact and the muscles stay attached to the same cuticle to make it possible for the insect to move. Towards the end of the molt just before ecdysis, there is secretion of certain proteases into the molting gel. Then, these enzymes work together in the digestion of the chitin and proteins in the old endocuticle down to their components: N-acetyl glucosamine sugars and amino acids. Further, for the production of the next cuticle, the reabsorption of molting fluid takes place into hemolymph which helps these components to be recycled. The process of reabsorption happens in two manners. Firstly, back through new cuticle and epidermis and secondly, through the gut via swallowing and uptake in hindgut.

Close to the end, where molting fluid is reabsorbed the insect undergoes ecdysis i.e., shedding of the old cuticle which takes place in a stereotyped arrangement of behavior. This behavior is characterized by a series of coordinated movements which helps to loosen the muscle attachments with the old cuticle. After this phase, there is a series of peristaltic waves which travel from posterior to anterior and helps to make the insect to rupture the old cuticle anteriorly and to free itself. The cuticle opens at ecdysial sutures. These are the areas of old cuticle which lack exocuticle. In order Lepidoptera, the head capsule has slipped down over the forming mandibles early in the molt to allow the formation of a larger head capsule. The old head capsule isolates from the rest of the old cuticle during ecdysis and falls off as the new larva leaves its old cuticle.

Dermal glands i.e., Verson's glands secretes a waterproofing cement layer on the top of epicuticle. When the larva sheds its old cuticle this layer sprawls over the surface as a result of larval movements under the old cuticle. There are certain cases in which there is a secretion of waxy layer on the top of this layer for the first few days after ecdysis which helps in preventing desiccation. The pore canals transversing through the cuticle from the epidermal cells help in secreting this waxy layer.

4.4 Post-ecdysial expansion and sclerotization

The larva fills its tracheae with air after the process of ecdysis and furthermore swallows air so as to expand the new larger cuticle. After achieving its final size, the new cuticle solidifies. It gets dark or tanned to changing degrees relying upon whether the cuticle is to be rigid or flexible.

Sclerotization is the phenomenon of hardening the exocuticle by cross-linking the proteins together with the chitin to form a balanced structure appropriate for an exoskeleton which helps to anchor the muscles to permit the movement. N-β-alanyldopamine and N-acetyldopamine are the two essential cross-linking agents. The N-β-alanyldopamine is found in tan cuticles of many pupae belonging to order Lepidoptera. The key enzymes for the formation of these compounds are phenoloxidase for conversion of tyrosine to dopa and dopa decarboxylase for conversion of dopa to dopamine.

4.5 Growth of the tissues

The form of the cuticle is determined by the epidermis which may grow either by an increase in cell number or by an expansion in cell size. During developmental period the cell number increases just before molting in larval stages of insects. For instance, the size of the larval forms of *Cyclorrhapha* increases with the increase in the size of epidermal cells.

The growth of central nervous system in hemimetabolous insects does not involve the production of new neurons except in the brain. In the terminal abdominal ganglion of *Acheta* Linnaeus, for example, there are about 2100 neurons at all stages of development. On the other hand, the number of glial cells in the ganglion increases from about 3400 in the first stage larva to 20,000 in the adult and the volume of the ganglion increases 40-fold. There is extensive reconstruction of the nervous system in holometabolous insects during metamorphosis and undifferentiated neuroblasts persist through larval period.

During larval development, the marked changes occur in the sensory system of hemimetabolous insects. At each molt additional mechanoreceptors and chemo-receptors are added to the already present receptors. The ommatidia forming the compound eyes also increase by number. In contrast, the number of sensilla remains constant throughout the larval life in holometabolous insects and compound eyes are only present in the adults.

The musculature in larval forms closely resembles to that of adults in most hemime-tabolous insects. In addition, there may be some muscles that are operative only during molting and later disappear after the final molt. During larval development, muscles increase in size and there is no basic change in their arrangement. The muscles grow by an increase in fiber size between molts and by the addition of new fibers at molts.

In case of epidermis, increase in the size of an internal organ results from an increase in cell size or in cell number or sometimes both. The increase in volume of internal structures especially the fat body is limited by the cuticle. In holome-tabolous insects, larvae with soft, folded cuticle, considerable growth is possible. The extension of the abdomen by unfolding inter-segmental membranes occurs in species with more rigid cuticles. In grasshoppers, and probably in some other insects, some growth of internal organs occurs at the expense of air sacs which become increasingly compressed during each developmental stage. The fat body of larval *Aedes* grows by an increase in cell number. In *Drosophila*, most of the tissues have constant number of cells and grow by cell enlargement. This enlargement is accompanied by endomitosis. In the midgut, both processes occur, the epithelial cells enlarge, but ultimately breakdown during secretion and each is replaced by two or more cells derived from the regenerative cells. In some insects the whole of the midgut epithelium is replaced at intervals by regenerative cells. In general, it appears that tissues which are destroyed at metamorphosis grow by cell enlarge-ment while those that persist in the adult grow by cell multiplication.

The development of Malpighian tubules varies. In Orthopteroid orders, Malpighian tubules increase in number throughout the larval life. The primary tubules arise as diverticula from the proctodeum in the embryo. There are four primary tubules in *Blatta*, some insects have six. Secondary tubules develop later, largely post-embryonically. For example, *Schistocerca* Stal has six primary tubules, but twelve more are added before the larva hatches and more develop in each larval stage up to the adult. Secondary tubules appear as buds at the beginning of each larval stage, but after their initial development they increase in length without further cell division as a result of an increase in cell size. In holometabolous insects, the number of Malpighian tubules remains constant throughout their larval life but tubules increase in length by increasing their cell size and by cell rearrangement.

5. Conclusion

Many animals possess a distinct immature developmental form (e.g.) in case of insects larval forms are distinct during developmental period. The immature forms are much more adapted to environmental conditions than adults and con-sume more food to undergo the process of transition from immature to adult form. Larval stages undergo metamorphosis in which they usually change in shape, size and organization to form an adult. These changes are triggered and monitored by hormones such as juvenile hormone. Class Insecta is characterized by four dif-ferent patterns of growth and development i.e., Ametabolous, Paurometabolous, Hemimetabolous and Holometabolous. Each pattern is characterized by specific morphological and hormonal changes. Insect larvae are broadly classified into four groups: Protopod larva, polypod larva, oligopod larva and apodous larva.

During the process of molting, the insect larvae molt with number of times. The number of instars varies across insect species. The environmental conditions like temperature, humidity, photoperiod along with other factors such as sex, inheritance, food quality and quantity affect the number of instars. Some insects can regenerate their lost appendages before the production of molting hormone. Regeneration can be seen in larval forms of Blattodea, Phasmatodea, some hemipterans and orthopterans. Larval development is also marked by significant changes in the sensory system in hemimetabolous insects. Mechanoreceptors, chemoreceptors are added along with the increase in size of muscles.

Larval development is a significant phase in the development history of an insect which molts and physiologically change the insect to adjust in different environmental conditions and habitats.

Author details

Amritpal Singh Kaleka*, Navkiran Kaur and Gaganpreet Kour Bali
Department of Zoology and Environmental Sciences, Punjabi University, Patiala, Punjab, India

*Address all correspondence to: apskaleka@gmail.com

IntechOpen

References

[1] Richards OW, Davies RG. Imms' general textbook of entomology. In: Imms AD, Richards OW, Davies RG, editors. Classification and Biology. 10th ed. Vol. 2. New York, USA: Chapman and Hall Halsted Press Book, John Wiley & Sons; 1977. pp. 433-443

[2] Jarjees EA, Merritt DJ. Development of *Trichogrammaaustralicum*Girault (Hymenoptera: Trichogrammatidae) in *Helicoverpa* (Lepidoptera: Noctuidae) host eggs. Australian Journal of Entomology. 2002;**41**:310-315

[3] Clifford HF, Hamilton H, Killins BA. Biology of the mayfly *Leptophlebiacupida* (Say) (Ephemeroptera: Leptophlebiidae). Canadian Journal of Zoology. 1979;**57**:1026-1045

[4] Sehnal F. Growth and life cycles. In: Kerkut GA, Gilbert LI, editors. Comprehensive Insect Physiology, Biochemistry and Pharmacology. Vol. 2. Pergamon, Oxford, United Kingdom: Postembryonic development; 1985. pp. 1-86

[5] Nijhout HF. Insect Hormones. Princeton, NJ: Princeton University Press; 1994

[6] Kamata N, Igarashi Y. Influence of rainfall on feeding behavior, growth, and mortality of larvae of the beech caterpillar, *Quadricalcariferapunctatella* (Motschulsky) (Lep., Notodontidae). Journal of Applied Entomology. 1994;**118**:347-353

[7] Telfer MG, Hassall M. Ecotypic differentiation in the grasshopper *Chorthippusbrunneus*: Life history varies in relation to climate. Oecologia. 1999;**121**:245-254

[8] Byers JR, Lafontaine JD. Biosystematics of the genus *Euxoa* (Lepidoptera: Noctuidae). XVI. Comparativebiology and experimental taxonomy of four subspecies of *Euxoacomosa*. Canadian Entomologist. 1982;**114**:551-565

[9] Westermann F. Wing polymorphism in *Capniabifrons* (Plecoptera: Capniidae). Aquatic Insects. 1993;**15**:135-140

[10] Leonard DE. Effects of starvation on behaviour, number of larval instars, and developmental rate of *Porthetriadispar*. Journal of Insect Physiology. 1970;**16**:25-31

[11] Roisin Y. Diversity and evolution of caste patterns. In: Abe T, Bignell DE, Higashi M, editors. Termites: Evolution, Sociality, Symbioses and Ecology. Dordrecht, The Netherlands: Kluwer Academic Publishers; 2000. pp. 95-119

[12] Cushman JH, Rashbrook VK, Beattie AJ. Assessing benefits to both participants in a lycaenid-ant association. Ecology. 1994;**75**:1031-1041

[13] D'Amico LJ, Davidowitz G, Nijhout HF. The developmental and physiological basis of body size evolution in an insect. Proceedings of the Royal Society of London. Series B. 2001;**268**:1589-1593

[14] Nijhout HF. Physiological control of molting in insects. American Zoologist. 1981;**21**:631-640

[15] Nijhout HF, Williams CM. Control of moulting and metamorphosis in the tobacco hornworm, *Manducasexta* (L.): Cessation of juvenile hormone secretion as a trigger for pupation. The Journal of Experimental Biology. 1974;**61**:493-501

[16] Rountree DB, Bollenbacher WE. The release of the prothoracocotropic hormone in the tobacco hornworm, *Manducasexta*, is controlled intrinsically by juvenile hormone. Experimental Biology. 1986;**120**:41-58

[17] Wang Z, Haunerland NH. Ultrastructural study of storage protein granules in fat body of the corn earworm *Heliothiszea*. Journal of Insect Physiology. 1991;**37**:353-363

[18] Gleiser RM, Urrutia J, Gorla DE. Body size variation of the floodwater mosquito *Aedesalbifasciatus* in central Argentina. Medical and Veterinary Entomology. 2000;**14**:38-43

[19] Fietz MJ, Concordet JP, Barbosa R, Johnson R, Krauss S, McMahon AP, et al. The hedgehog gene family in *Drosophila* and vertebrate development. Development. 1994;**1994**:43-51

[20] Vincent JFV, Wegst UGK. Design and mechanical properties of insect cuticle. Arthropod Structure & Development. 2004;**33**:187-199

Chapter 4

Edible Insects Diversity and Their Importance in Cameroon

Meutchieye Félix

Abstract

Insects are known to be part of Sub-Saharan African region. Entomophagy is a common practice in Cameroon food systems. The current chapter is based on both original research and major literature review in the domain. A variety of insects species and consumable stages, as well as preference and their spatial distribution are presented in this chapter. Insects are described according to the recent taxonomy features and their bioecology is provided. Some consumption patterns, preferences and determinants are described. The role of insects consumption is also highlighted as well as some prospective investigation targeting edible insects preservation and sustainability in Cameroon. The paper points out some policy gaps that need to be addressed to harness the potentials of edible insects in Cameroon food systems.

Keywords: entomophagy, ecology, food systems, nutrition, conservation

1. Introduction

Most of sustainable development goals are food related and should be addressed properly. Particularly in Africa, food and nutrition insecurity is coupled to a growing demand of animal source proteins that could be solved by a different conception of food systems [1]. The ideas about considering insect production as part of livestock have been a long run scholar discussion in the West [1, 2] and emerged as a consistent topic in Africa of recent [3, 4]. Insect consumption also called entomophagy is then widely accepted as a palliative to food scarcity not only for today but also for the future [1, 5, 6]. Insect consumption is part of cultural heritage in tropics and beyond [6–8]. Insects at different stages are nutritive food sources made cheaper by their availability and sustainable by their nature [5, 7, 9]. Cameroon has been regularly cited for its richness of edible insects and related practices [3, 4]. The objective of this review is to summarize the major features concerning edible insects, few constraints and opportunities for the country in a globalized and changing environment.

2. Distribution of entomophagy in Cameroon

Cameroon is an elongated country stretching from Congo Basin (humid tropics) to Lake Chad (Sahel). The country is also described as Africa in miniature by its diversity and position. The entomophagy is present everywhere from Sahelian to humid forest regions [3–5]. There is a variety of beliefs attached to insect consumption all over the country [5, 10]. Some communities have a larger panel of edible insects throughout the year than other [11]. The differences observed concerning practices and recipes among communities are based on preferences, ability to harvest and process as well as the social importance attached to insect consumption [11, 12]. There is a large number of insects consumed in Cameroon.

3. Diversity of edible insects

Compared to other food sources, there are little taboos in insects' consumption in most of Sub-Saharan countries [3]. About 1700 insects' species are consumed around the world with majority from tropical world [1]. Some are widely identified and few less known in the context of recent interest by research in Cameroon.

3.1 Major groups

Caterpillars (order Lepidoptera) are the most populated group of edible insects in Cameroon concerning the number of species [10, 11, 13]. About 200 species have been reported and many are still unveiled. Some are used by limited communities, making their identification a real challenge [5, 11]. Forests provide shelter and perfect milieu for numerous of them [5, 7, 9]. Populations have developed a great knowledge about the bioecology of the most marketed species. In savannah and less woody forests, there are less diversity like in Adamawa and northern regions [10]. **Figure 1** below displays the naturel gathering nest (in a tree) and fresh and dried caterpillars found in Cameroon.

The second group of most consumed insects is made up of termites (order Blattodea) with two species, *Macrotermes* sp., being the most exploited [10–12, 14]. Termites are widely exploited all over the country [10]. The common recipe is to grill them after removing wings with onions, raw or not, as shown in **Figure 2** [14].

Besides "soft" species, crunchy and crispy are very regarded as a delicacy, mainly of Coleoptera order [15]. Adult Palm weevils are known as pest but "domesticated" by raffia palm harvesters (adults and grubs) [10, 11]. Various adults' species are the most collected in *Cetonia* sp. (**Figure 3**) [13], larvae are the most demanded is palm weevil, particularly the yellowish skin which is grown in raffia palm (see **Figure 4**) in Cameroon household [11]. Some tree, shrubs or humus grubs are also exploited in particular communities and need to be properly identified [10, 13].

The acridians or grasshoppers (Orthoptera), *Locusta migratoria* (also *Schistocerca gregaria*) known as migratory locusts are commonly exploited in drier regions of Cameroon (**Figure 5**), whereas *Ruspolia differens*, *Tettigonia viridissima* and *Zonocerus variegatus* are exploited in humid environments in southern Cameroon [10, 14]. *Zonocerus* sp. is treated with care because of toxins if not well prepared before ingestion.

Figure 1.
Edible caterpillar nest and heap of fresh and dried ones (Imbrasia sp.).

Figure 2.
Processing of termites (Macrotermes sp.) and selling packages in Cameroon.

Figure 3.
Adult dried cetonia (Cetonia aurata) in Cameroon Western highlands.

Figure 4.
Palm weevil grubs (Rynchophorus sp.) and a dish made with in Cameroon.

Figure 5.
Migratory locusts (greyish and greenish) and dried in North Cameroon.

Field crickets and cricket-like species (order Orthoptera) are also exploited in Cameroon humid regions [10, 11]. Giant cricket also called tobacco cricket (*Brachytrupes membranaceus*) is widely harvested. Some species are believed to have particular uses in pharmacopoeia beside being considered as food.

3.2 Specialties

Insects have played important roles in some national communities. For instance, some forest groups like Baka people praise the taste, flavors, contents and properties of some insects at particular stages [5, 11]. Honey bees larvae are considered as a powerful detoxification agent and then prescribed to some recovering patients in those communities [11]. In the same line, red ants (*Oecophylla longinoda*) are used by some forest communities in Centre region to cure manhood impotency. These biting red ants are collected, grilled and mixed with some other local herbs and fruits before consumption [11].

4. Edible insects' trade

Insects contribute also to income generation in rural and peri urban households [5, 11]. Local and transboundary trade involving Cameroon has been mentioned in some studies [5, 14]. Insects' harvesting requires some abilities acquired by empirical experiences which made some people experts and providing livelihoods means. This is for instance the case of *Guizigua* and *Mofu* men reputed to master termites' ethology [16]. The rapid urbanization is African towns is also promoting agro-industries, with impact on insects' consumption [17]. There is a growing economy of edible insects though the primary objective of insects' collection is own consumption [10, 11, 14]. As shown by **Figure 6**, there are many reasons attracting consumers to insects' utilization as food sources, with varying weight.

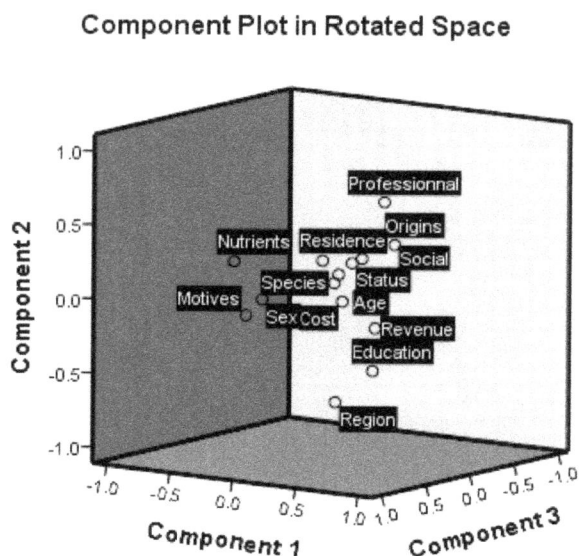

Figure 6.
Component rotational analysis of insects' consumption determinants.

4.1 Marketing channels, actors and benefits

Edible insects have become a more or less "normal" item or commodity in local trading spaces, even in big cities [17, 18]. The main categories of actors involved in collection are youths (irrespective of sex) for seasonal insects, women and youths for crickets and locusts, men for grubs and termites [5, 10, 11, 18]. Intermediate categories are made mainly by women or youths (street vendors) and exclusively women for urban markets. In *Mofu* and *Baka* communities some insects are exclusively harvested by men under specific conditions [12, 16]. Benefits are determined by the quality of products (freshness, color, origin and presentation), particularly in town markets [5, 11, 18].

4.2 New trends

With the influence of the mainstream food systems, processing and labeling edible insects are sectors for new food ventures [15, 17]. The type and flavors of spices make the difference for some conserved or already processed items [10]. Fast communication is now harnessing the pickup of technologies and making edible insects' as a potential popularized food item [1, 2, 6].

5. From harvesting to farming

Insect production for human consumption started very humbly under a group of researchers from Gembloux—Belgium [2]. Preliminary findings in artificial production of palm weevil larvae are promising [19]. Massive palm grub production is in pilot stages with some important success [19–23]. Substrates and microclimate conditions have been mastered and trials already done under farmers' conditions [22]. The investment for edible insects' production is balanced with substantial income attached and also to environmental health.

6. Some aspects of nutrition and food safety

If supposed nutritional factors with food habits have factored much entomophagy in Cameroon like in other places, some field experiences are giving evidence [1, 6, 19, 24]. Science communication and access are making consumers more attracted by appealing presentation of edible insects' nutrition facts. Palm weevil larvae has been largely studies [25] and results are opening avenues for other larvae consumption [26]. Entomophagy is then no more a fancy or traditional habit, but a real response to nutritional insecurity [27, 28]. Food safety in edible insects encompasses both endogenous and environmental factors. Chemical analyses have shown that some insects need special care before their consumption, like *Zonocerus variegatus*, because of its gut contents [29]. Under agrochemical pollution, natural harvesting is no safer as highlighted by some consumers in Cameroon suffering of indigestion [11, 24]. Pesticides are contaminants becoming a potential problem for consumers of in Cameroon western highlands [30].

7. Perspectives about policies and research gaps

Insects' consumption even popular or popularized is not yet legal in Cameroon context [30] because of insufficient policies. Most of gathered food in nature in Cameroon is not yet under strict safety regulation which may bring serious public health issues in case of hazardous products. While insects' farming and trade are gaining population attention, research and other influencers should in connivance with other international bodies invest in the domain. Vocational and higher education systems should also embrace genuinely the trends by crafting appropriate curricula and learn from learning by doing experiences. *Mofu* community has a wide range of knowledge on insects in their environment [31]. Indigenous knowledge built for centuries could speak a lot to research and policy today for better future of better human nutrition security, adapted to the context of biodiversity depletion [32]. From old, new can grow, new roads from within to many other places where collaboration are opened. The International Centre of Insect Physiology and Ecology (ICIPE) is already theorizing interesting and promising channels, and some in Cameroon about insects' domestication.

8. Conclusion

Insects' consumption in Cameroon is a reality for ages. Its biodiversity is very rich and massively exploited diversely. There are large numbers of species, exploited for food and other reasons. Urbanization and communication facilities coupled to food systems unification are favoring the migration of habits across generations and regions. Besides nutritive values at home level, edible insects provide income, jobs and potential enterprises, both for economic, scientific and cultural impacts in Cameroon society. There are gaps that make excellent avenues for innovative approaches in insects research. On the food safety aspects, risks analyses regarding natural or secondary toxicity of consumed insects should be undertaken. Edible insects' value chain could be made also stronger is legal framework is well designed and implemented. There are needs to train Veterinarians and others on harvesting methods as well as processing and packaging, which could yield norms. These could mainstream numerous affordable insects within animal proteins food systems and improve nutrition security and income generation at many extents. Some indigenous habits and skills (related to products' properties and bioecology) should be well studied and not regarded as lower knowledge. Many communities practices could be excellent avenues for future, wider than food security alone. Edible insects' research and policy need to be strengthened locally with the existing findings already uncovered.

Acknowledgements

We are thankful to authors, namely, Dr. Le Gall, Dr. Muafor and M. Gnetegha for providing their research findings and/or comments to blend the current synthesis.

Conflict of interest

The author declares to have not "conflict of interest."

Author details

Meutchieye Félix
Faculty of Agronomy and Agricultural Sciences, Department of Animal
Production, University of Dschang, Dschang, Cameroon

*Address all correspondence to: fmeutchieye@gmail.com

IntechOpen

References

[1] van Huis A, Van Itterbeeck J, Klunder H, Mertens E, Halloran A, Muir G, et al. Edible Insects: Future Prospects for Food and Feed Security. Fao Forestry Paper 171. Rome: FAO. 2013. p. 188

[2] Hardouin J. Production d'insectes à des fins économiques ou alimentaires: Mini-élevage et BEDIM. Notes fauniques de Gembloux. 2003;**50**:15-25

[3] Niassy S, Affognon HD, Fiaboe KKM, Akutse KS, Tanga CM, Ekesi S. Some key elements on entomophagy in Africa: Culture, gender and belief. Journal of Insects as Food and Feed. 2016;**2**(3): 139-144. DOI: 10.3920/JIFF2015.0084

[4] Niassy S, Ekesi S. Contribution to the knowledge of entomophagy in Africa. Journal of Insects as Food and Feed. 2016;**2**(3):137-138. DOI: 10.3920/JIFF2016.x003

[5] Balinga MP, Mapunzu PM, Moussa J-P, N'gasse G. Contribution des insectes de la forêt à la sécurité alimentaire, l'exemple des chenilles d'Afrique centrale. In: Produits forestiers non ligneux Document de Travail No. 1; FAO. 2004. p. 107. Available from: http://www.fao.org/forestry/site/6367/en

[6] Bize V. Les insectes, une resource alimentaire d'avenir? Insectes. 1997;**3**(106):11-13

[7] Vantomme P. Les insectes forestiers comestibles, un apport protéique négligé. Unasylva. 2010;**61**(236):19-21

[8] Schabel HG. Forests insects as food: A global review. In: Proceedings of a Workshop on Asia-Pacific Resources and Their Potential for Development (Forest Insects as Food: Humans Bite Back); 19-21 February 2008; Chiang Mai, Thailand. 2010. pp. 37-64

[9] Johnson DV. The contribution of edible forest insects to human nutrition and to forest management: Current status and future potential. In: Proceedings of a Workshop on Asia-Pacific Resources and Their Potential for Development (Forest Insects as Food: Humans Bite Back); 19-21 February 2008; Chiang Mai, Thailand. 2010. pp. 5-22

[10] Miantsa FO, Meutchieye F, Tanebang C. Diversité et Répartition des Insectes Comestibles au Cameroun. Saarbrucken: Editions Universitaires Européennes; 2016. p. 56. ISSN: 978-3-639-48385-7

[11] Meutchieye F, Tsafo KE, Niassy S. Inventory of edible insects and their harvesting methods in the Cameroon Centre region. Journal of Insects as Food and Feed. 2016;**2**(3):145-152. DOI: 10.3920/JIFF2015.0082

[12] Cloutier J, editor. Insectes Comestibles en Afrique: Introduction à la Collecte, au Mode de Préparation et à la Consommation des Insectes (Agrodok 54). 1st ed. Wageningen: Fondation Agromisa et CTA; 2015. p. 81. ISBN Agromisa: 978-90-8573-147-4; ISBN CTA: 978-92-9081-578-5

[13] Muafor FJ, Angwafo TS, Le Gall P. Biodiversité des insectes de la ligne volcanique du Cameroun: Distribution altitudinale d'une famille de Coléoptères. Entomologie faunistique—Faunistic Entomology. 2011;**63**(3):195-197

[14] Tsafo KEC, Meutchieye F, Manjeli Y. Exploitation alimentaire et nutritionnelle des insectes comestibles en zone forestière du Centre Cameroun. Sciences naturelles et agronomie. 2016;**2**:245-254. ISSN 1011-6028

[15] Muafor FJ. Des insectes au menu: Les scarabées, friandises croquantes. Sciences au Sud—Le journal de l'IRD. 2012;**63**:8

[16] Seignobos C. Des insectes au menu: Les maîtres des termites. Sciences au Sud—Le journal de l'IRD. 2012;**63**:8

[17] Tabuna H. Des insectes au menu: De l'économie de ramassage à l'agro-industrie. Sciences au Sud—Le journal de l'IRD. 2012;**63**:8

[18] Meutchieye F, Niassy S. Preliminary observations on the commercialisation of *Rynchophorus phoenicis* larvae at Mvog-Mbi market in Yaoundé, Cameroon. Journal of Insects as Food and Feed. 2016;**2**(3):199-202. DOI: 10.3920/JIFF2015.0081

[19] Muafor FJ, Gnetegha AA, Le Gall P, Levang P. Palm weevil farming contributing to food security in sub-Saharan Africa. In: van Huis A, Tomberli JK, editors. Insects as Food and Feed, from Production to Consumption. Wageningen Academic Publishers: Wageningen; 2017. DOI: 10.3920/JIFF2018.x003 73

[20] Ebenebe CI, Okpoko VO, Ufele AN, Amobi MI. Survivability, growth performance and nutrient composition of the African palm weevil (*Rhyncophorus phoenicis* Fabricius) reared on four different substrates. Journal of Bioscience and Biotechnology Discovery. 2017;**2**:1-9. Article Number: JBBD-28.09.16-018. Available from: www.integrityresjournals.org/jbbd/index.html

[21] Quayea B, Atuaheneb CC, Donkohc A, Adjeid BM, Opokue O, Amankrahf MA. Alternative feed resource for growing African palm weevil (*Rhynchophorus phoenicis*) larvae in commercial production. American Scientific Research Journal of Engineering, Technology, and Sciences (ASRJETS). 2018;**48**(1):36-44

[22] Fogang Mba AR, Kansci G, Viau M, Ribourg L, Muafor FJ, Hafnaoui N, et al. Growing conditions and morphotypes of African palm weevil (*Rhynchophorus phoenicis*) larvae influence their

lipophilic nutrient but not their amino acid compositions. Journal of Food Composition and Analysis. 2018;**69**: 87-97. DOI: 10.1016/j.jfca.2018.02.012

[23] Monzenga Lokela JC, Le Goff GJ, Kayisu K, Hance T. Influence of substrates on the rearing success of *Rhynchophorus phoenicis* (Fabricius). African Journal of Food Science and Technology. 2017;**8**(1):7-13. DOI: 10.14303/ajfst.2015.065

[24] Meutchieye F, Tsafo KE, Mekongo Fombod UC, Niassy S. La consommation d'insectes comme stratégie d'atténuation de la malnutrition: Expériences de la région du Centre Cameroun. AGRIDAPE Revue sur l'Agriculture durable à faibles apports externes. 2014;**30**(4):13-14

[25] Fogang Mba AR, Kansci G, Viau M, Hafnaoui N, Meyneier A, Demmano G, et al. Lipid and amino acid profiles support the potential of *Rhynchophorus phoenicis* larvae for human nutrition. Journal of Food Composition and Analysis. 2017;**60**:64-73. DOI: 10.1016/j.jfca.2017.03.016

[26] Dounias E. Des insectes au menu: Vous prendrez bien une petite larve? Sciences au Sud—Le journal de l'IRD. 2012;**63**:9

[27] Le Gall P. Des insectes au menu: Entomophagie en Afrique. Sciences au Sud—Le journal de l'IRD. 2012;**63**:7

[28] Muafor FJ, Gnetegha AA, Le Gall P, Levang P. Exploitation, Trade and Farming of Palm Weevil Grubs in Cameroon. Working Paper 178. Bogor, Indonesia: CIFOR; 2015. DOI: 10.17528/cifor/005626

[29] Ehigie LO, Okonji RE, Ehigie AF. Biochemical properties of thiaminase, a toxic enzyme in the gut of grasshoppers (*Zonocerus variegatus* Linn). Cameroon Journal of Experimental Biology. 2013;**9**(1):9-16. DOI: 10.4314/cajeb.v9i1.2

[30] Miantsia FO, Meutchieye F, Niassy S. Relationship between new farming practices and chemical use and the consumption of giant cricket (*Brachytrupes membranaceus* Drury, 1770). Journal of Insects as Food and Feed. 2018;**4**(4):295-300. DOI: 10.3920/ JIFF2018.0010. ISSN 2352-4588 Online

[31] Seignobos C, Deguine J-P, Aberlenc H-P. Les Mofu et leurs insectes. Journal d'Agriculture Traditionnelle et de Botanique Appliquée. 1996;**36**(2):125-187

[32] Niassy S, Fiaboe KKM, Affognon HD, Akutse KS, Tanga CM, Ekesi S. African indigenous knowledge on edible insects to guide research and policy. Journal of Insects as Food and Feed. 2016;**2**(3):161-170. DOI: 10.3920/ JIFF2015.0085

Chapter 5

Econometrics of Domestication of the African Palm Weevil (*Rhynchophorus phoenicis* F.) Production as Small-Scale Business in Ghana

Thomas Commander N., Jacob P. Anankware,
Onwugbuta O. Royal and Daniel Obeng-Ofori

Abstract

A reconnaissance survey of the domestication of the African palm weevil (APW) (*Rhynchophorus phoenicis*), which produces the edible larvae that are cherished as a delicacy among many tribes in Ghana, was conducted. Out of a total number of 560 semi-trained farmers, 271 (48.39%) were actively engaged in *R. phoenicis* farming near their homes or gardens, while 289 (51.61%) were non-active. Economic viability analyses showed that the active farmers would break even and repay their loans of GH¢1000 when they produce 3020 larvae at unit selling price of GH¢0.33, within a period of 4 months and 7 days (17 weeks). In a year, a farmer would have three production cycles and generate a total revenue of GH¢3018.79, at average monthly production of 755 edible larvae, net cash availability of GH¢1448.79, and projected net profit of GH¢448.79 in the first year of production. The farmer would make more profit and become wealthy in business in subsequent years. The pilot scheme of palm weevil farming was viable and ameliorated poverty and malnutrition of rural farmers in Ghana.

Keywords: African palm weevil (*Rhynchophorus phoenicis*), larvae, domestication, farming, revenue

1. Introduction

The practice of eating insects as food by humans is human entomophagy [1]. Although this is traced to the biblical literature, nevertheless, eating of insects by humans had been a taboo in many western nations [2]. Insect farming and entomophagy has developed fast to emerge as a strong economic contributor in recent times. It is estimated that insect eating is practiced regularly by over 2 billion people worldwide [3]. In most tropical countries like Nigeria, Ghana, Cameroon, Congo, Angola, and South Africa, the utilization of insect species as food has been reported [4]. The world population is expected to be above 9.2 billion people, adding more than 2 billion to an already crowded planet earth in 2050 [5].

The Food and Agriculture Organization (FAO) estimates about 1.2 billion are suffering from chronic hunger globally. Therefore, the major challenge is achieving global food security including the hungry populace by 2050 [5]. Meeting this massive additional demand for food will require concerted actions on a number of fronts including focusing on increasing the production and consumption of underutilized and underappreciated natural food resources such as insects [3]. Edible insects constitute about 80% of the animal kingdom [6–8]. This has high potentials of contributing to food and nutrition security globally. Another aspect is to increase the promotion and consumption of other foods which have been constrained by production; processing and trade laws and challenges are to be considered. Although human entomophagy is widely reviled in European and North American societies, more than 1900 species of insects have been reported as edible by humans in over 112 countries of the world, particularly in Asia, the pacific, Africa, Latin America, and Europe [9]. However, the consumption of different species of the edible larvae of palm weevils (***Rhynchophorus*** spp.) has been confirmed as highly nutritious globally. According to the World Health Organization (WHO), every 100 g of palm weevil larvae contains 182 kcal of energy, 6.1% of protein, 3.1% of fat, 9.0% of carbohydrates, 4.3 mg of iron, 461 mg of calcium, and other important vitamins and minerals [10]. A reconnaissance survey of the semi-cultivation (domestication) of African palm weevil (*Rhynchophorus phoenicis*) assessed the profitability of domesticating the production of edible larvae near homes and gardens in Ghana and was undertaken in September to October, 2016. Semi-cultivation refers to the domestication of the breeding of African palm weevil (APW) near living homes and gardens for proper management and continuous production of the edible larvae all year round. This step has been taken to avert the risk of sourcing for the edible larvae from the host plants, raffia palm (*Raphia hookeri*) and oil palm (*Elaeis guineensis* Jacq.), from the wild, since it has been difficult to successfully mass-rear the edible larvae under laboratory conditions, for sustainable production for consumption and commercialization of the edible larvae throughout the year [11] instead of its sensational availability in Ghana and other parts of the world [1].

2. Materials and method

2.1 Study area

The study was conducted in two regions (Ashanti Region and Brong-Ahafo Region) in Ghana. Four communities, Bomfa, Asotwe, Doyina, and Amoafo-Bekwai, were visited in Ashanti Region, while one community, Jema, was visited in the Brong-Ahafo Region. The selection of these communities was based on accessibility and availability of the host plants, oil palms (*Elaeis guineensis*) and raffia palms (*Raphia hookeri*), in Ghana.

2.2 Data collection and training of farmers

Out of a total of 560 trainees, 406 from Ashanti Region and 154 farmers from Brong-Ahafo were engaged and trained in a palm weevil larvae production. The farmers were interviewed orally and salient data were taken. Questionnaires were sent to farmers in some localities which could not be directly accessed. On the process of domestication, the rural farmers were taught how to collect adult palm weevils from the infested host plants (oil and raffia palms).

(a)   (b)

Figure 1.
(a) Adult (female) of R. phoenicis and (b) edible larvae of R. phoenicis.

2.3 Breeding activities

The breeding of the palm weevils was carried out in laboratory following known methods [12, 13]. The breeding took place in plastic containers where sex identification, mating, and oviposition activities were carried out under suitable ambient environmental conditions. The captured adult palm weevils were fed with portions of the natural host plants (raffia or oil palm stems) which were renewed on weekly basis to enable the young larvae develop to food size within 3–4 weeks and harvested for sale to consumers. Some final instar larvae were allowed to develop to pupal stages and emerged as adults to ensure continuity of adult generations. Upon completion of the training, each farmer was given a seed capital of GH¢1000 (which was equivalent to US$250) to procure the needed infrastructure and commence production by the Aspire Food Group (AFG), an international nongovernmental organization (NGO) that initiated the pilot scheme in Ghana. Profitability of the business was calculated using break-even point analysis, while cash flow statement was obtained by direct method [14]. **Figure 1a** and **b** shows the adult palm weevil and the larvae, respectively.

3. Results

A total of 560 farmers were trained on palm weevil production across 5 communities in 2 regions of Ghana. Out of which, 271 (48.39%) were active farmers, while 289 (51.61%) were non-active farmers (**Table 1**). **Table 2a** and **b** showed the various items that constituted the fixed and variable costs estimated at GH¢285 and GH¢715, respectively. The total number of 500 larvae produced within a cycle of production at the unit selling price of GH¢0.3332 (**Table 3**) yielded total sales of GH¢166.6. **Table 4** showed the first break-even point of an active farmer who produced 3020 edible larvae and sold them to consumers to raise funds to repay the loan of GH¢1000. The number of three production cycles which an active palm

S/No.	Regions	Communities	Number trained	Active farmers	Non-active farmers
1	Southern Ghana Ashanti Region	Bomfa	120	20	100
		Asotwe	70	65	5
		Doyina	200	87	113
		Amoafo-Bekwai	70	45	25
2	Brong-Ahafo Region	Jema	100	54	46
		Total trained farmers	560	271	289
		Percent	100	48.39	51.51

Table 1.
Number of trained farmers in palm weevil larvae production in Ghana.

Expenses (items)	Units	Amount per unit	Sub total GH¢	Total
(a) Fixed cost				
Nursery/shield	1	450	450	
Plastic containers	10	10	100	
Machete	1	35	35	
Watering can	3	10	30	
Labor/construction of shield nursery	—	—	100	
Total fixed cost (TFC)			715	715
(b) Variable cost				
Palm trunk (sup/feed)			100	
Local transport per day (21 days)			126	
Miscellaneous/sundry			59	
Total variable cost (TVC)			285	285
Grand total cost (TFC + TVC)				1000

Table 2.
Cost of palm weevil production.

No. of larvae per cycle harvest (21 days) (GH¢)	Unit selling price (GH¢)	Total sales (GH¢)
1	0.3332	0.3332
(370–500 yield) 500	0.3322	166.6

Table 3.
Income estimate for a cycle of production.

weevil farmer would carry out successfully within a year was estimated to generate a revenue of GH¢3018.79 (**Table 5**). The net cash available for the total investment within the farming year after repayment of the loan value of GH¢1000 was estimated to be GH¢1448.79 (**Table 6**), while the summary of the annual financial report showed a net profit of GH¢44.879 with first farming year (**Table 7**).

Items	Variables	Outcomes
Selling price (SP)	166.6	—
Variable cost (VC)	285	
Contribution (SP-VC)	—	−118.4
Contribution (C)	118.4	—
Number of larvae (NOL)	500	—
Contribution/larvae (^{c}/NOL)	—	0.2368
Fixed cost (FC)	715	
Contribution/larvae (C/L)	0.2368	
Break-even point nos larvae (FC/CL)	—	3020 larvae
Break-even point nos of larvae (BEP)	3020	
Unit selling price (USP)	0.3332	
Break-even point sales (BEP larvae × USP)	—	GH¢ 1006.26

Table 4.
Break-even point in palm weevil farming.

Items	Variables	Outcomes
Break-even point larvae (BEP larvae)	3020	—
Number of larvae (Nos)	500	
Number of cycles to break even (BEP/Nos)	—	3.0
Number of cycles to break even	3.0	—
Number of days in a cycle	21	—
Number of days to break even	—	126.84
Number of months to break even	—	4 months 2 days
Number of months in a year	12	
Larvae per month (3020/4)	—	755
Annual larvae production (12 × 755)	—	9060
Annual revenue (9060 × 0.3332)	—	GH¢ 3018.79

Table 5.
Production cycles/annual revenue.

	Unit cost (GH¢)	Total cost (GH¢)
Operating activities		
Cash from customers	3018.79	
Operating cost to be paid (285 × 3)	(855)	
Net cash flow from operating activities	—	2163.79
Investing activities		
Purchase/construction of fixed assets	(715)	
Net cash flow from investing activities		(715)
Financing activities		
Loan received	1000	

	Unit cost (GH¢)	Total cost (GH¢)
Loan repayment	(1000)	
Net cash flow from financing activities	—	
Net increase in cash/cash equivalent		1448.79
Net projected profit		448.79

Table 6.
Cash flow statement for the first year.

Items	Expenses (GH¢)	Revenue (GH¢)
Total revenue		3018.79
Less cost		
Operating cost	855.00	
Fixed cost	715.00	
Loan repayment	1000.00	
Total cost	2570.00	
Net profit		448.79

Table 7.
Summary of annual financial report of palm weevil farming.

4. Discussion

The study has shown that palm weevil farming is profitable and a thriving small-scale business in four communities (Bomfa, Asotwe, Doyina and Amoafo-Bekwai) in the Ashanti Region and Jema community in the Brong-Ahafo Region of southern Ghana, where the indigenes were vested with traditional cultivation and harvesting of the larvae from host plants in the wild [15]. The survey showed that over 48.39% (271) farmers who benefited from the palm weevil larvae breeding training were able to establish, manage, and harvest their larvae at least once within 3–4 weeks, while 51.61% (289) farmers were inactive and could not manage their palm weevil farms successfully. This is an indication that acquiring training for weevil larvae production requires low level of formal education of the trainee. This also agrees with earlier reports which stated that income earning from rearing and processing of edible insects is generally at subsistent level in Cameroon and other African countries [16]. All the actively trained farmers who secured a loan of GH¢1000 were able to successfully establish a palm weevil farm and achieved a break-even point within 4 months and 7 days (17 weeks). In a cycle of production, each farmer could raise a total number of about 3020 larvae at unit selling price of GH¢0.332 translating to a total revenue of GH¢1006.26. In all, a farmer will have three production cycles and generate a total annual revenue of GH¢3018.79 and make a net profit of over GH¢448.79 in the first year of operation and make greater profit in subsequent years. Palm weevil larvae are a good source of protein [17] and contain polyunsaturated fatty acids such as omega-3 and omega-6 which are recommended for diabetic and hypertensive patients with heart disorders [18, 19]. Therefore, the harvested larvae formed reliable alternative source of high-quality animal protein food for the local communities for healthy living [20]. The breeding of palm weevil larvae also provides employment opportunities for youths, women, and aged pensioners who have retired from the civil service [15]. Considering the high socioeconomic benefits accruing to the farmers, it is recommendable that the

palm weevil larvae farming be adopted as veritable scheme for economic empower-
ment of rural farmers in tropical African countries. However, the scheme needs to
be upgraded and supported with more funds by donors to make greater impact by
training more people and increasing the seed capital for those successfully trained
to commence the business. Consequently, there is a need to provide funds to support critical aspects of researches that will lead to mass production of the edible
larvae of the African palm weevil which have been found to be a "super food" that
provides acceptable amounts of "macro and micro" nutrients required for enhancing the longevity of life. Palm weevil production has been a small-scale family-
based business which can translate to a Cooperative Association of Insect Producers
to attract funding from financial institutions and influence policy-making toward
their trade and development in rural communities in Africa as advocated by the
Food and Agriculture Organization (FAO).

5. Conclusion

The domestication of palm weevil currently practiced at small-holder level in
Ghana is profitable. It is a source of additional income stream of the families, thus
enhancing the financial capacity of the farmers and improving their socioeconomic
status. Therefore, international and national agencies of government and nongov-
ernmental organizations (NGOs) associated with poverty alleviation, malnutrition,
and food security should fund research and development activities, especially
that funding of the commercial rearing of this important natural food resource is
inevitable in Africa and, indeed, all over the world.

Author details

Thomas Commander N.[1], Jacob P. Anankware[2*], Onwugbuta O. Royal[3]
and Daniel Obeng-Ofori[4]

1 Department of Biological Sciences, Niger Delta University, Bayelsa State, Nigeria

2 Department of Horticulture and Crop Production, School of Agriculture and
Technology, Sunyani, Ghana

3 George and George Chartered Accountants, Port Harcourt, Nigeria

4 Office of the Vice Chancellor, Catholic University College of Ghana, Sunyani,
Ghana

*Address all correspondence to: anankware@yahoo.com

IntechOpen

References

[1] Anankware JP, Osekre AE, Obeng-Ofori D, Canute KM. Identification and classification of common edible insects in Ghana. International Journal of Entomology Research. 2016;**1**(5):33-39

[2] Thomas CN, Okwakpam BA, Ogbalu OK, Empere CE. Utilisation of the larvae of *Rhynchophorus phoenicis* F. (Coleoptera: Curculionidae) as human food in Niger Delta, Nigeria, Niger Delta. Biologia. 2006;**6**(2):18-22

[3] Van Huis A, Van Iterbeeck J, Klunder H, Mertens E, Halloran A, Muir G, et al. Edible Insects: Future Prospects for Food and Feed Security. Rome: Food and Agriculture Organization of the United Nations (FAO); 2013

[4] Ntukuyoh AI, Udiong DS, Ikpe E, Akpakpan AE. Evaluation of nutritional values of termites (*Macrotermes bellicosus*): Soldiers, workers and queen in Niger Delta Region of Nigeria. International Journal of Food Nutrition and Safety. 2012, 2012;**1**(2):60-65

[5] FAO. The State of Food and Agriculture. Rome: Food and Agriculture Organization of the United Nations; 2009

[6] Victor YL. Basic Invertebrate Zoology. P.M.B.1515 Ilorin, Nigeria: Ilorin University Press; 1988

[7] Premalatha M, Abbasi T, Abbasi SA. Energy efficient food production to reduce global warming and ecodegradation: The use of edible insects. Renewable and Sustainable Energy Review. 2011;**15**:4357-4360

[8] Alamu OT, Amao AO, Nwokedi CI, Oke OA, Lawal IO. Diversity and nutritional status of edible insects in Nigeria: A review. International Journal of Biodiversity and Conservation. 2013;**5**(4):215-222

[9] Durst PB, Shono K. Edible forest insects: Exploring new horizons and traditional practices. In: Proceedings of a Workshop on Asia- Pacific Resources and Their Potential for Development. 19-20 February 2008. Chiang Mai, Thailand, Bangkok: Food & Agricultural Organization of United Nations; 2010. pp. 1-4

[10] Mercer CWL. Sustainable production of insects for food and income by New Guinea villagers. Ecology of Food and Nutrition. 1997;**36**(2-4):151-157

[11] Ebenebe CI, Okpoko VO. Preliminary studies on alternative substrates for multiplication of African palm weevil under captive management. Journal of Insects for Food and Feed. 2016;**2**(3):171-177

[12] Thomas CN. Biology utilisation and roaring of African palm weevil (*Rhynchophorus phoenicis* F) in palms of the Niger Delta, Nigeria [PhD Thesis]. Nkpolu, Port Harcourt, Nigeria: Department of Biological Sciences, Rivers State University; 2003

[13] Hoddle SM. Entomophagy: Faming Palm Weevil. Riverside, CA, USA: Food University of California; 2013

[14] Adeniji AA. An Insight: Management Accounting. 4th ed. Lagos, Nigeria: El-Toda Ventures Ltd; 2008. p. 93

[15] Offenberg J, Wiwatwitaya D. Weaver ants (*Oecophylla smaragdina*) convert pest insects into food: Prospects for the rural poor. In: Paper presented at the International Conference on Research, Food Security, Natural Resource Management and Rural Development,

University of Hamburg, Germany, 6-8 October 2009. 2009

[16] Muafor FJ, Gnetegha AA, Le Gall P, Levang P. Exploitation, trade and farming of palm weevil grubs in Cameroun. In: Working Paper 178. Bogor, Indonesia: (CIFOR) Centre for International Forestry Research; 2015

[17] Ayieko MA, Oriaro V. Consumption, indigenous knowledge and cultural values of the Lakefly species within the Lake Victoria region. African Journal of Environmental Science and Technology. 2008;**2**(10):282-286

[18] Finke MD. Complete nutrient composition of commercially raised invertebrates used as food for insectivores. Zoo Biology. 2002;**21**:269-285

[19] USDA. National Nutrient Database for Standard References. 2012. Available at: www.ars.usda.gov/ba/bhnre/ndi. [Accessed: 12-12-12]

[20] Okaraonye CC, Ikewuchi JC. *Rhynchophorus phoenicis* (F) larvae meal: Nutritional value and health implications. Journal of Biological Sciences. 2008;**8**:1221-1225

An Insect Bad for Agriculture but Good for Human Consumption: The Case of *Rhynchophorus palmarum*: A Social Science Perspective

Rafael Cartay, Vladimir Dimitrov and Michael Feldman

Abstract

This article presents a review of the current state of the art in the study of human consumption of insects in the Amazon basin and, in particular, of the larva of the beetle *Rhynchophorus palmarum* which is the insect of greatest consumption by the native indigenous communities of the Amazon basin. It includes detailed information on cultivation, collection and consumption, as well as the dietary, medicinal and symbolic role the *Rhynchophorus* plays in a variety of Amazonian cultures. The article emphasizes aspects related to its role as vector of a plague that damages commercial agriculture of palms and some fruit trees, as opposed to its role as a food source that constitutes a rich source of protein of high biological value.

Keywords: edible insects, Amazonian protein, insect's nutritional value, *Rhynchophorus palmarum*, Amazonian indigenous diet

1. Introduction

Insects have attracted the attention of mankind since ancient times for both negative and positive reasons. Negative, related to their destructive effects on agricultural and industrial crops, causing large economic losses, and their harmful effects on human health, causing huge human losses by transmitting diseases such as Chagas disease, dengue, malaria, yellow fever, chikungunya, leishmaniasis and others. Positive, related to their use as a human food source, of particular importance to help mitigate, in the medium term, critical cases of food insecurity and famine, and feeding other animal organisms [1]. Insects play a key role as regulatory elements of terrestrial ecosystems, fundamental in pollination processes, important as predictors and bioindicators of environmental changes [2] and to evaluate the impacts of fragmentation of plant cover, fire and invasive plants [3, 4]. Insects are also used as bioindicators of plant stress [5], elements to enrich the soil [6], accelerate the recycling of detritus [7] and for the biological control of pests [2, 8, 9]. In many cultures they are useful as effective popular medicines [10–14], and cutting edge medical technology [15]. Insects are highly valued, in many parts of the world, as symbols in religious rituals and in other cultural practices [16–21].

The importance of insects is remarkable from a multidimensional perspective related to human culture [22], and especially in relation with biodiversity. Insects represent the animal group with the most evolutionary success [22]. They also constitute the largest animal biomass on the planet [23], with a higher volume than the rest of the animals together ([24], pp. 67–68). Insects have the advantages of abundance (wide geographical distribution and great adaptability), productive facility (high reproduction rate, easy handling and cultivation, efficiency in food conversion and great potential for internal and external commercialization) [13, 24–26], and a high nutritional value suitable for human and animal uses [13, 25, 27–31]. Insects are, for these reasons, an excellent food alternative for a world with a growing human population, which lives in a scenario characterized by an inequitable distribution of productive land, employment and income, and which faces serious problems in accessing enough quality food for expanding populations [11, 30, 32–41].

Around the world, more than 1 million species of insects have been described by science, while the existence of 5–10 million more is estimated, yet to be described [42], which makes them the group of animals of the greatest diversity on the planet. Of the total described, there are, according to the most conservative estimates, between 1900 [37, 43] and 2000 species of insects [11], used as food by nearly 3000 ethnic groups in more than 102 countries [11, 24].

Considering the relationship between the number of edible insect species with respect to the total number of insect species, we find that only 0.2% of the described species are edible, which represents just 0.033% of the total estimate of insect species, described or not. Of the total number of insects, nearly 60,000 described species live in the Amazon basin. There, the proportion of edible insect species, estimated at about 135 species, gives a figure of 0.00225% with respect to the total of regional insect species. This means that the percentage of edible insect species in the world is negligible (0.2%, of the total described, and 0.033% of the estimated total), and even more so in the case of the Amazon (0.00225%).

When examining the taxonomic concentration of the insect species described and, in particular, edible insects, we find that, approximately, 74%corresponds to four orders: *Coleoptera* (35%), *Diptera* (15%), *Hymenoptera* (12%) and *Lepidoptera* (12%) [44, 45]. There are some insects more consumed than others, individually, such as certain species of crickets, grasshoppers and locusts, while the most consumed in Amazonia are the larvae of the beetles *Rhynchophorus palmarum* and *Rhinostomus barbirothis* [27, 46, 70]. That preference in consumption varies according to areas, and there are notable exceptions. In the Brazilian Amazon, the largest portion of the vast Amazon basin, there are about 135 species of edible insects. Among them, the most consumed species belong to the order of hymenoptera, which include ants, termites, wasps and bees, especially excelling in the consumption of ant species *Atta cephalotes* and *A. sexdens*. The same happens in other smaller areas of the basin, such as the south of the Colombian Amazon, where some indigenous groups like the Andoque, who live in the middle part of the Caquetá River, are notable consumers of parasol ants, of the genus *Atta* [47].

When taking into account the fidelity level of insect consumption in the Amazon basin, i.e., the frequency of their use as a dietary component, it is observed that only 30 species of insects are frequently consumed, highlighting, among them, the consumption of *Rhynchophorus palmarum* (*Rp*, hereinafter). Paoletti et al. [46] reported a consumption of 6 kg/year/per capita of *Rp* larvae. As each larva in its fresh state weighs between 8 and 12 g, it would imply the consumption of 50 larvae per person per month, which is possible. Ramos-Elourdoy and Viejo Montesinos [24] point out that the Yanomami indigenous group consumes more than that, in addition to other insects (ants, wasps and other larvae) and spider, which is not, strictly speaking, an insect [48]. Beckerman [49] reported similar consumption among the Bari of Venezuela. In summary, it can be concluded at this point that, although a large percentage of indigenous insects are not consumed in the Amazon basin, there is a high consumption of some species, such as *Rp*, which appears as a supplement to the diet in many Amazonian indigenous communities [50], together with medicinal uses [51].

The objective of this article is to review the double impact of the larva of *Rhynchophorus palmarum* (*Rp*), both in its destructive effect on cash crops causing significant economic losses, and from the perspective of the valuable benefits it provides to the Amazonian indigenous communities by supplementing their diet, especially during times when there is a shortage of hunting and fishing production.

2. Methodology

To collect the information needed for this research, which is part of a larger investigation, two methods were used. First, the method of in situ observation, carried out directly in a number of native indigenous communities of the Peruvian Amazon, supplemented by informal interviews with members of these communities, particularly those located near the cities of Iquitos and Nauta, in the Loreto region, during the period from May to July 2015. Several popular regional markets were visited, and especially the large market of Bethlehem, to interview small traders, some informal, who regularly offered products derived from about 20 varieties of Amazon palms (parts of the plant: drupe, palmetto or inflorescence of the bud, and related insects). This field work included an excursion for the collection of edible insects (in particular Suri, *Rp*), guided by young people of the Yagua ethnic group, in the Nanay river basin.

The second method consisted of a neat bibliographic-bibliographical review carried out in two specialized libraries located in the cities of Iquitos: the rich library of the Institute of Amazonian Studies of the Peruvian Amazon (IIAP) and the beautiful library of the Center for Theological Studies of the Amazon (CETA) In addition, information was collected over several months in libraries in Lima, particularly the one from the French Institute of Andean Studies (IFEA) and the Institute of Peruvian Studies (IEP). During that time we also interviewed personalities linked to different aspects of Amazonian life: the historian and novelist Róger Rumrrill, the journalist and novelist Juan Ochoa-López, the chefs Pedro Miguel Schiaffino and Pilar Agnini and the anthropologist Alberto Chirif, one of the greatest experts and analysts of the Peruvian Amazon from the perspective of the social sciences.

3. Results

The approach to the subject of the investigation can be considered in three parts. In the first we describe the *Rp*, and especially the preferred edible larval state. In the second part we describe the behavior of the *Rp Coleoptera* as a pest, and in particular as a vector of a nematode that causes serious economic losses to commercial agriculture, most notably in the cultivation of African palm and coconut palm, as well as some fruit trees. In the third part we address the topic of *Rp* as an edible insect of importance in the diet of the Amazonian indigenous groups, and as an alternative to contribute, in the medium and long term, to a solution of the serious problems of food insecurity confronting a growing population, without regular access to an abundance of other protein-rich foods and that confronts notable food shortages now and possible catastrophic shortages in the future.

3.1 Description of the *Rp* and its larval stage (form)

The *Coleoptera Rp* belongs to the order *Coleoptera*, family *Curculionidae*, tribe *Rhynchophorini* (see **Figure 1**). The genus *Rhynchophorus* is made up of 10 species. Of these, three are present in the neotropics: *R. cruentatus*, *R. richeri* and *R. palmarum*. The *Rp* is a widely distributed species in the Neotropics, from southeast California and Texas to Bolivia, Peru, Paraguay, Uruguay and Argentina, in an altitudinal range from 0 to 1200 m above sea level [52–54].

The *Rp coleoptera* is known by many common names: cucarrón, cigarrón, weevil, palm weevil, casanga, black weevil, and coconut palm weevil. Its larva is called, in the Amazonian regions, Suri (Peru), Chontacuro (Ecuador), Gualpa (Colombia), Palm Worm (Venezuela), apart from the many other names it is given in different parts of the Amazon basin: mojojoi, mojomoi, mojotoi, casanga, mukint, mujin, and headworm.

It is a matt black beetle, with a size that varies between 2 and 5 cm. In adult state, this coleopter presents sexual diformism, that is, the male is different from the female. The female has the beak curved and smooth, and longer than that of the male. The male is easily recognized because, in addition, he carries a tuft of mushrooms in the dorsal part of the beak. Both male and female show activity both in the day and in the night: they are observed in the fallen trunks of the palms during the early hours of the morning or at the end of the afternoon, although they are more active towards 11 o'clock at night ([55–57], pp. 11–13).

The female lays her cream white eggs, of a size that fluctuates between 1.0 and 2.5 mm, in palm trunks. It deposits them, in an average of 900 units, in vertical position on the soft tissue of the open trunk of the palm, protecting it with a brown waxy substance. After 2–4 days, the larvae emerge, without legs and with a body length of a little more than 3 mm, slightly curved in the belly. From there it begins its development in nine instars, which last between 42 and 62 days, until it reaches instar IX, when it becomes a pupa. It then takes 30–45 days for the adult to emerge, and from 7 to 11 days to leave the cocoon [55, 58].

The females oviposit in the cuts of the petiolar bases of the palms with wounds or rot. There, inside the infected palm, usually near the rotting bud, the insect develops, fulfilling its total life cycle until reaching its final form ([58], p. 21), depending on the material or substrate on which it feeds (colonized substrate). The life cycle ranges from 119 to 231 days, when they are raised in the laboratory [59], and under normal conditions, a minimum of 122 days: 3.5 days as eggs, 60.5 days as larvae, 16 days as a nymph and 42 days as an adult [24, 60, 61]. The females have an oviposition period of up to 43 days. A female can oviposit up to 63 eggs in a day, and

Figure 1.
The final form is the edible white larval stage.

from 697 to 924 during her entire cycle [55, 62–64]. In the final instar, the larva has a length of 5–6 cm, and a weight of 12–30 g [65].

The *Rp* females are attracted by the volatile compounds that emanate from the palms with wounds or rot, seeking to feed on their soft tissues. Thirty-one species of *Rp* host plants have been registered, belonging to 12 families. Among them, the *Palmaceae* family predominates with 19 species, mainly *Elaies guineensis* and *Cocos nucifera*, of great economic importance. Of the 19, there are 11 species of Amazonian palms host of the *Rp*. Among them, *Mauritia flexuosa, Maximiliana regia, Bactris gasipaes, Oenocarpus bataua, Euterpe oleracea, Astrocaryum huicungo*, of great importance for human nutrition in the Amazon basin. Of the 11, 3 species of palms are very affected: aguaje, morete, muriti or moriche (*Mauritia flexuosa*), ungurahui, ungurahua or seje (*Oenocarpus bataua*) and cucurito (*Maximiliana regia*) [66]. *Rp*, a polyphagous insect, also causes damage to fruit trees such as papaya, mango, avocado, orange, guava, by feeding on ripe fruits. And, in addition, on sugarcane, banana, cacao and pineapple. But there is a difference: in these plants, *Rp* produces damage, but does not behave like a pest. It acts like this only in the case of palms and sugar cane [55, 54].

3.2 *Rp* as a plague

Rp is a devastating plague affecting some palms of economic importance that constitute commercial plantations such as coconut and oil palms, and of some Amazonian palms of great utilitarian interest for native indigenous communities ([67], pp. 151–156). When *Rp* is attracted to the wounds and rotting in the stems and the bud of the palms, it deposits its eggs in the soft tissues and the tree is infected by the nematode *Bursaphelenchus cocophilus* (*Bc*, hereinafter), which is the main cause of ring syndrome, known as red or small leaf, which has devastated the coconut and African palm plantations located in Central and South America.

The *Bc* nematode is an obligate migratory endoparasite, which lives all of its life inside the palm and without multiplying inside the disseminating insects [55]. The nematode is acquired by the *Rp* larva, which acts as its main vector, maintaining it through the molts until reaching the adult stage. By leaving the diseased palm, it can infect three or four healthy neighboring palms. The combat and control campaign is currently done using traps or plastic containers (olfactory scent traps), placing pheromones of synthetic or natural origin to attract the insects. The traps are placed in the field at a distance of 1–2 hectares in the most infected areas [68, 69].

3.3 *Rp* as food

In the case of the Amazon basin, the larvae of the *Rp* and *Rhinostomus barbirothis* beetles are the most consumed [27, 70], although the primacy corresponds, with a great advantage, to *Rp* [46]. It should be noted, however, that this statement is not generalizable for all countries in the basin. A very notable exception is Brazil, in whose Amazonian region mainly hymenoptera insects (ants, termites, wasps and bees) are consumed ([11], p. 423; [47]).

Rp larvae are a source of proteins and fats used in native Amazonian indigenous communities to supplement their diet, under normal conditions based on hunting, fishing and farming. This source of protein could also play a larger role in the diet in times of need, as the larva *Rp* constitutes, as do edible insects in general, a protein possibility of high biological value and low cost. It is interesting to note that in urban areas of many Amazonian regions, edible insects are freely available. In the Iquitos markets, *Rp* larvae are sold in different presentations: live, cooked and roasted. Vargas et al. ([71], p. 65) pointed out that an average of 3500 units are sold there per day, especially on weekends.

Some researchers several decades ago posed the need to value the consumption of insects as an excellent food resource, widely used among Amazonian Indians, among Mexican rural dwellers and in many Asian and African cultures. These authors [15, 72] considered that protein malnutrition among indigenous groups in the Amazon was relatively low in the area due to its high consumption of insects, fungi, drupes and almonds. That opinion, perhaps a bit exaggerated, can be sustained with some reservations. Riparian natives satisfy their protein needs basically with the consumption of fish. Some riparian groups have an average per capita consumption of 20–50 kg per year, although in some communities they reach consumption levels close to 200 kg per year. In these conditions, the consumption of insects plays a secondary role, complementing the diet, not as a primary component but as a necessary complement.

Just as insect consumption has been overestimated in some studies, so in others such consumption has been underestimated. Many times indigenous people do not recognize this consumption in the food consumption surveys that are applied to them. The Indians in the most advanced process of cultural assimilation do not declare that consumption because they have learned in the cities that this consumption is considered unpleasant and dirty. This concealment does not occur with indigenous groups that are proud of their ethnic identity and boast of such a food practice.

Although not declared openly, the consumption of insects is common throughout the Amazon basin. That is evident if one makes a visit to any indigenous community. In some native communities of the Loreto region, in the Peruvian Amazon, we directly recorded the consumption of nine species of insects belonging to several orders, although the most consumed was the *Rp*, in close correspondence with the wide geographic distribution and the abundance of some host plants such as *Mauritia flexuosa*, known as aguaje, because in the low jungle there are huge stands of that palm that are known as aguajales ([67], p. 160). The indigenous inhabitants consume the larvae fresh, alive, or dead, roasted or fried. In the urban areas of the Amazon they are served fried in their own fat, or roasted over a direct fire. This form of preparation and consumption constitutes an imitation of traditional indigenous preparation. The most sophisticated urban chefs offer their product in salad, or wrapped in the manner of a Tequeño, or roasting the larvae on a skewer as if it were a Turkish kebab. This is the case in the restaurants of Iquitos, in Peru, or Puyo, in Ecuador, or Leticia, in Colombia, or in Puerto Ayacucho, in Venezuela. Other cooks have incorporated the larva into some typical preparations of the regional cuisine of the Peruvian Amazon. We thus have the juane de chonta (palmito), which mixes palmita, tender edible inflorescence of some Amazonian palms, with suris or palm worms [73]. Brewer-Carías ([74], p. 150) recommends cutting the posterior end of the larva before consuming it raw, to reduce its spicy flavor, probably caused by its digestive juices.

Depending on the season, and rising or falling river levels, which changes the availability of food in the jungle by affecting the relative productivity of hunting and fishing, recollection is used to augment these primary sources. This activity includes wild fruits, drupes of palm trees, fungi, mollusks, small terrestrial animals such as amphibians, and edible insects. The consumption of these insects is very important during some times of the year. Paoletti et al. [46], based on studies by Ramos-Elourdoy and Viejo Montesinos [24], using various sources, recorded much higher consumption among the Yanomami, an indigenous group that inhabits the Venezuela-Brazil border, during particular seasons.

The nutritional value of edible insects is sufficiently proven by numerous laboratory studies. The protein content of edible insects varies between 30 and 40%: from 30% for wood larva to 80% in the wasp *Polybia* sp. [75], which equals, and even exceeds, the values obtained for different types of meat, typically ranging between 40 and 75%. In the specific case of the larva *Rp*, the protein content is 76%, clearly higher than that of beef, which is 50–57%. Something similar can be observed in the fat content of protein sources. In meats this ranges from 17% in fish, to 19% in beef. In the case of the larva of the coleoptera, this value oscillates between 21 and 54%, presenting, in addition, a better composition. The skin, in particular, is rich in oils. These are fatty oils of the unsaturated type: linoleic, linolenic and other polyunsaturated fats [31, 39, 43, 65, 71, 76–78]. Regarding the total caloric value, the *Coleopterous* larvae have caloric values around 560 kcal/100 g, higher than the 430 kcal/100 g of beef [11].

The protein of animal origin is important for its high biological value, which depends on the number and variety of essential amino acids in its content, and its digestibility or ease of assimilation by the human body. The biological value of the protein corresponds to the proportion of protein absorbed and used by the organism. To be used most efficiently, protein is required to have all the essential amino acids in the right proportions. This happens with foods of animal origin such as milk and meat. The protein of edible insects is also of high biological value, similar to that of meats, with a triple advantage over them: it has a lower relative price, is easier to digest and is healthier because it does not have cholesterol [79]. In addition, if a protein of high biological value, such as insects or meat, is consumed and combined with another of lower biological value, such as cassava or plantain, the foods complement each other, and the biological value of the resulting dishes increases. However, the consumption of insects' greatest importance for the conservation of the environment lies in the fact that it has a better efficiency index for the conversion of food into biomass.

The value of an animal as a source of nutrients depends mainly on its nutritional contribution and on the efficiency with which this animal converts the food consumed into biomass [75, 80]. In this respect, the animal that gains the most weight for each gram of food consumed is more efficient. To obtain 1 kg of beef, 13 kg of food is needed. For chicken, the most efficient among the commonly consumed animals, 6 kg is required. On the other hand, only 2 kg are needed for insects, showing a high rate of conversion efficiency. For Costa-Neto [81] and Krajick [75], insects are more efficient in relative terms than other animals, because they are invertebrate, cold-blooded animals. The disadvantage they present is that their consumption is seasonal and their production is not currently significant in terms of volume sufficient to supply the potential market. This situation can be reversed, and we are beginning to see large-scale cultivation in some countries of the world, such as Thailand, Mexico and Spain.

Recollection of insects is an activity carried out by the far majority of Amazonian indigenous communities. To analyze its cultural dynamics and dietary contribution, it is necessary to understand the changes produced in the larva. The timing of the *Rp* instar stages are important to determine the period of collection in the jungle, behavior that the natives know perfectly and transmit as ancestral knowledge, ethnoetology as it is called by Posey [48]. To be collected, the instar must be at least 1 week old since its incubation period, to ensure that it is a viable larvae, that it has reached a major stage in its evolution, with a weight close to 12 g, that it is as fat as a finger and has a color between cream and brown. Guzmán-Mendoza [18] points out that it is important to distinguish the onset of the larval stage in order to consider a larvae as food. The ideal conditions occur after about 2 months of life, after the period of infestation has occurred, which does not always occur immediately after the cutting of the palm. There are periods more favorable than others to infest the downed logs. Ramos-Elourdoy et al. [82] point out that the period of greatest infestation occurs, in the Amazon basin, between August and October, depending on the environmental zones and the rainy season, during which stem rot accelerates, making the trunk softer and propitiating the perforation by the insect. This can occur naturally or be induced by humans.

When human action intervenes, the collecting activity goes much further, becoming in practice a work of cultivation or protoculture, as Ramos-Elourdoy and Viejo Montesinos [24] called it. In this case, the indigenous person fells the palm, and in the downed trunk makes an incision approximately 10 cm × 10 cm, leaving a mark to identify the place. Two months later he or she returns to the site, knowing what to look for and where to look [83]. The collector comes back this time with an ax and a container to collect the larvae. He or she then opens the bark of the trunk with the ax, and extracts 30–40 larvae each time, part of the harvest of a day. A whole palm tree can produce over 500 larvae. Then, the collector takes the larvae home to consume with the family. The period of greatest collection goes, in the Peruvian Amazon, from July to October, both in the Lower and Upper Amazonian basin. Depending on the season, the identity of the collectors changes. If it is a hunting or fishing season, and men are absent from the community, women and children are responsible for the collection.

The "cultivation" or "proto-culture" of *Rp* is not a simple task. Araujo and Becerra [27] and Arango-Gutiérrez [84] point out that the Yekuana and Piaroa ethnic groups, from the Venezuelan Amazon, induce and promote the breeding of *Rp*, a highly esteemed insect in their culture, to which they attribute great food virtues. They collect the larvae from the fallen palm trunks and transport them to their homes, where they are fed with pieces of soft plant tissues from selected palm trunks. For this purpose they prefer tissues of the seje palm (*Jessenia bataua*), arguing that, when consumed, they give the larvae a better flavor. For the "cultivation" of the *Rp*, they intentionally chop healthy palms, section their trunks longitudinally to attract and concentrate a greater number of infesting individuals on the food source, favoring copulation and oviposition. After a lapse of 35–45 days, they harvest the larvae and consume them, simmering them until they are crispy. Bukkens [76] notes that the collection is planned and highly predictable relative to initiating the infestation of the downed trunk.

Neto and Ramos-Elourdoy ([11], p. 430) point out that the collection of edible insects depends on four factors: food restrictions and taboos, traditional customs, personal taste or taste preference of the group and the search for food security to guarantee survival. Another factor could be added: seasonality, because in the rainy season the process of insect infestation is accelerated. These authors also argue, together with Miller [85], that the use of an insect as food is related to four variables: the environment, the availability and accessibility of insects, the mode of production and the forms of reproduction of the insect and culture and food restrictions. In several native communities the collection and cultivation of insects corresponds to indigenous women and children, and they exhibit a festive spirit while accomplishing this task, which they perform even in times of abundance of hunting and fishing products.

3.4 *Rp* as a symbol

It is known that when consuming food, symbols, meanings, and signifiers are consumed at the same time. In such a way, the consumption of insects goes beyond obtaining nutrients in moments of scarcity or to supply deficiencies of proteins and fats. Every food substance must be viewed from a three-dimensional perspective, because it provides, at the same time, nutrients, medicines and symbols. The consumption of insects in the different regions of the Amazon basin is inscribed within a culture, whose members use symbols to communicate, as individuals and as a social entity, and to express them and think about their culture. Foods contain messages or stories that serve, along with other cultural elements, to insert themselves into the worldview or matrix of a culture. These messages are transmitted inter-generationally and incorporated, with adjustments, into the dietary patterns of a social group [86]. An excellent illustration of the symbolic consumption of insects is the study carried out by Acuña-Cors [87] in an indigenous community of the Reyes Metzantla, in Puebla, Mexico. Also notable are the mentions made by Macera and Casanto ([88], p. 242) of the symbolism associated with the suri larva (*Rp*) among the communities of the Ashaninka indigenous Amazonian ethnic group. For its members, the larva suri (imooqui) has an owner or tutelary god, the Imoobo, which must be asked for permission before consuming the insect. In the Asháninka legend, the Imoobo is an old woman who ends up being addicted to eating so much suri. Jara ([47], p. 226), meanwhile, it is said that the Andoque and Desana, indigenous peoples of the Colombian Amazon, large consumers of *Rp*, see in the metamorphosis of the insect the expression of a transforming magical power. The beetle, regarded as the father of the larva, is attributed a male generative power, which penetrates with its beak, the symbolic phallus, the perforated trunk of the palm, which corresponds to the vagina. The larva is, for them, a hybrid animal/vegetable product that is produced within a process of shamanistic transformation. Mexican indigenous groups, such as the Mazahua, in the state of Mexico, consider insects as mediators between the earthly and the supernatural worlds. Using them, the Indians raise supplications to God asking him, for example, to send them rain ([44], p. 85). The Yucuna, from the Colombian Amazon, distinguish three types of *Rp* beetle larvae, which they call mojojoi: the mumuna, small; the huachurú, the median, and ñamaja, the largest ones. Here the larvae are collected by women and children, and are subject to barter or gifting. Giving them is a demonstration of affection, which they deliver by wrapping them in pieces of palm leaves and tying them with vine fiber ([83], pp. 83).

The symbolism of insect consumption is different when it occurs among non-Amazon urban consumers. In this context, the edible insect leaves everyday life to become an exotic matter that, in some cases, produces amazement, and can become an object of consumption and gastronomic tourism. However, in most cases consumption of insect is viewed with horror by visitors from other cultures, who consider insects dirty, disease-ridden pests, and which arouses feelings of apprehension and disgust, which can even cause phobias and neuroses and even physical illness.

4. Discussion

The subject of edible insects has been attractive for popular magazines, but not as much for scientific research. Even in Latin America, where insects are consumed in almost all countries, there are still a lot of reservations about the matter, as if it were an exotic food practice exclusive to the most backward and unimportant indigenous communities. There have been few researchers who address this area of study, the exceptions being mostly European and American investigators. Among Latin Americans, researchers from Brazil and Mexico stand out, and some have made great contributions in the field [89]. Most of the studies done in Mexico are devoted to the study of insects grouped in *Coleoptera* and *Lepidoptera* [90], while research on insects from the Amazonian basin focuses on insects belonging to the groups of the *Coleoptera* and *Hymenoptera*, using the methodological support offered by applied ethnology [89]. Throughout the Amazon basin, the larvae most consumed are the *Rp* larvae ([27], of which few monographs have been written in relation to their abundance, leaving some areas untouched. There is a lack of comparative studies of the nutritional values of edible insects, as well as the biological value of the proteins they provide. It would also be interesting to understand the way indigenous groups put together their nomenclature, their classification systems and the specifics of consumption of such insects [91]. Little is known about the medicinal uses of these insects, which Ramos-Elourdoy [92] so emphatically raises, with the name of "nutracéutical entomofauna". We are just beginning to study the optimal manner to "cultivate" them in the jungle and to "raise" them domestically. Little is known about the efforts being made in Mexico, Thailand or Spain to foster large-scale insect production, in order to meet current and potential regional and global demand.

On the symbolic aspects of the consumption of insects, and the comparative cultural representations between the different ethnic groups of Central America and Mexico, as well as of South America, there is much work to do. Little research has been done on insect consumption between the populations of the Caribbean islands and black African-American populations. The elaboration of didactic manuals is necessary to develop popular enterprises related to the "cultivation" of edible insects. The specialists in the culinary arts must write recipes that introduce novel ways to facilitate the consumption of insects, overcoming the reservations that people have concerning their consumption. On the subject, only entomologists, ethnozoologists and applied anthropologists have been concerned thus far, but not nutritionists, for whom it should be a major concern. They, and various health organization, have the difficult task of developing efficient campaigns, attractive from the point of view of "taste", to promote the consumption of insects of high nutritional value, as an effective tool in reducing the serious problems of chronic malnutrition that affects a large percentage of the child population of developing world, constituting a situation of food insecurity that impacts the political, socioeconomic and public health realities in these countries. This study tries, at a minimum, to be a critical revision of the current state of the art around this topic, but it leaves many unanswered questions that must be approached by other investigators interested in the subject.

5. Conclusions

Edible insects represent an important source of protein and fats in the diet of the indigenous Amazonian population, particularly during times when the availability of products derived from hunting and fishing is reduced, these being the main and usual sources of necessary proteins. The protein derived from insects is of high biological value, due to its excellent content of essential amino acids, both in variety and quantity. It also results in easy digestibility, a relatively low energy cost and a high efficiency index in feed/biomass conversion. These attributes make the consumption of edible insects an attractive alternative that could be used, in the medium term, to tackle the serious problems of chronic malnutrition worldwide, if the adequate measures were taken to promote its large-scale production and consumption.

The *Rp*, in particular its larva, is the insect most consumed by native indigenous communities throughout the Amazon basin. In many cases it is more than a product that is the object of simple collection, because its "cultivation" is induced by the indigenous population, using ancestral knowledge and proven techniques. Its consumption, within an indigenous society, acquires a broad and deep connotation: as food, medicine and symbol.

Despite its abundance and importance in the diet of indigenous people of the Amazon, as an alternative source of protein and fats, the *Rp* has been little studied by the members of the South American academia and, in general, by Latin American experts, except for the pioneering studies of a few Mexicans and

Brazilians. Most of the current contributions on the subject come from American and European researchers. However, for a decade, interest in edible insects has been increasing among researchers from the countries that make up the Amazon basin, especially from Colombia, Ecuador, Venezuela and Peru, focusing mainly on aspects related to the nutritional value of the insect. There is, however, a lot to investigate in relation to the socio-cultural aspects involved in the use of edible insects in the Amazon, as well as the feasibility of larger scale production and consumption in areas without a cultural history of insect consumption.

Author details

Rafael Cartay[1,3], Vladimir Dimitrov[1,2*] and Michael Feldman[1,4]

1 Research Department, Clarivate Analytics Ecuador, Manta, Manabí, Ecuador

2 Institute of Organic Chemistry with Centre of Phytochemistry (IOCCP), Bulgarian Academy of Sciences, Sofia, Bulgaria

3 Research Department, Univ Tecn Manabi, Portoviejo, Manabí, Ecuador

4 Boston University, Boston, Massachusetts, USA

*Address all correspondence to: vladimir.dimitrov@investigador.ec

IntechOpen

References

[1] Iannacone J, Alvariño L. Diversidad de la artropofauna terrestre en la Reserva Nacional de Junín, Perú. Ecología Aplicada. 2006;**5**(1-2):171-174

[2] Pacheco V, Solari S, Velazco PM. A new species of Carollia (*Chiroptera: Phillostomidae*) from The Andes of Peru and Bolivia. Occasional Papers; Museum of Texas Tech University 236, 2004. pp. 1-15

[3] Gove AD, Majer JD, Rico-Gray V. Ant assemblages in isolated trees and more sensitive to species loss and replacement than their woodland counterparts. Basic and Applied Ecology. 2009;**10**:187-195

[4] Underwood EC, Fisher BL. The role of ants in conservation monitoring: If, when, and how. Biological Conservation. 2006;**132**(2):166-182

[5] Saha BC, Nichols NN, Cotta MA. Ethanol production from wheat straw by recombinant *Escherichia coli* strain FBR5 at high solid loading. Bioresource Technology. 2011;**102**(23):10892-10897

[6] Fortanelli Martínez J, Servín Montoya ME. Desechos de hormiga arriera (*Atta mexicana* Smith), un abono orgánico para la producción hortícola. Terra Latinoamericana. 2002;**20**(2):153-160

[7] Lamure JP, Martínez ML. El impacto de productos veterinarios sobre insectos coprófagos: Consecuencias sobre la degradación del estiércol en pastizales. Acta Sociológica Mexicana. 2005;**21**:137-148

[8] Schowalter TD. Insect Ecology an Ecosystem Approach. 3rd ed. New York: Academic Press; 2000

[9] Valadares LCA, Pasa MC. Use of plants and animals by the riverine population from Rio Vermelho Community, central western Brazil. Interações (Campo Grande). 2012;**13**(2):225-232

[10] Neto EMC, Pacheco JM. Utilização medicinal de insetos no povoado de Pedra Branca, Santa Terezinha, Bahia, Brasil. Biotemas. 2005;**18**(1):113-133

[11] Neto EC, Ramos-Elorduy J. Los insectos comestibles de Brasil: etnicidad, diversidad e importancia en la alimentación. Boletín Sociedad Entomológica Aragonesa. 2006;**38**:423-442

[12] Serrano-González R, Guerrero-Martínez F, Pichardo-Barreiro Y, Serrano-Velázquez R. Los artrópodos medicinales en tres fuentes novohispanas del siglo XVI. Etnobiología. 2013;**11**(2):23-34

[13] Vantomme P. Los insectos forestales comestibles, una fuente de proteínas que se suele pasar por alto. Unasylva. 2010;**236**(61):19-21

[14] De La Cruz LE, Gómez y GB, Sánchez CMS, Junghans C, Martínez JLV. Insectos útiles entre los tsotsiles del Municipio de San Andrés Larrainzar, Chiapas, México. Etnobiología. 2015;**13**(2):72-84

[15] Ratcliffe BC. (1990). The significance of scaraf beetles in the ethnentomology of non-industrial, indigenous people, 159-185. Posey, D.A./ W.L. Overal. Etnobiology: Implications and Applications. Belém: MPEG.

[16] Cano-Contreras EJ, Martínez MC, Balboa ACC. La abeja de monte (Insecta: Apidae, Meliponini) de los choles de Tacotalpa, Tabasco: conocimiento local presente y future. Etnobiología. 2013;**11**(2):47-57

[17] Márquez LJ. Meloponiculltura en México. Dugesiana. 1994;**1**(1):3-12

[18] Guzmán-Mendoza R, Herrera-Fuentes MDC, Castañc-Meneses G, Zavala-Hurtado JA, León-Cortés JL. La hiperdiversidad de los insectos: explorando su valor biológico, cultural y económico. In: Navarrete-Heredia JL, Castaño-Meneses G, Quiroz-Rocha GA, editors. Facetas de la Ciencia: Ensayos sobre Entomología Cultural. Guadalajara: Universidad de Guadalajara; 2011. pp. 51-54

[19] Lara Vázquez JA, Villeda Callejas MDP. Odonatos en la manifestación cultural de los pueblos. Revista Chapingo. Serie Ciencias Forestales y del Ambiente. 2002;8(2):119-124

[20] Pasa MC, Valadares LCA. Pest control methods used by riverine from Rio Vermelho community, South of Mato Grosso State, Brazil. Biodiversidade. 2011;9(1):4-14

[21] Viesca González F, Romero Contreras A. La Entomofagia en México. Algunos aspectos culturales. El Periplo Sustentable. 2009;(16):57-83

[22] Guzmán-Mendoza R, Calzontzi-Marín J, Salas-Araiza MD, Martínez-Yáñez R. La riqueza biológica de los insectos: análisis de su importancia multidimensional. Acta Zoologica Mexicana. 2016;32(3):370-379

[23] Aragón GA, Mcrón MA, López-Olguin JF, Cervantes-Peredo LM. Ciclo de vida y conducta de adultos de cinco especies de Phyllophora Harris, 1827 (Coléoptera: Mololonthidae). Acta Zoologica Mexicana. 2005;21(2):87-99

[24] Ramos-Elorduy J, Viejo Montesinos JL. Los insectos como alimento humano: Breve ensayo sobre la entomofagia, con especial referencia a México. Boletín Real Sociedad Española de Historia Natural. Sección Biología. 2007;102(1-4):61-84

[25] Rumpold BA, Schlüter OK. Nutritional composition and safety aspects of edible insects. Molecular Nutrition & Food Research. 2013;57(5):802-823

[26] Sayyed AH, Cerda H, Wright DJ. Could Bt transgenic crops have nutritionally favourable effects on resistant insects? Ecology Letters. 2003;6(3):167-169

[27] Araujo Y, Becerra P. Diversidad de invertebrados consumidos por las etnias Yanomami y Yekuana del Alto Orinoco, Venezuela. Interciencia. 2007;32(5):318-323

[28] Paoletti MG, Buscardo E, Dufour DL. Edible invertebrates among Amazonian Indians: A critical review of disappearing knowledge. Environment, Development and Sustainability. 2000;2(3-4):195-225

[29] Pijoan M. El consume de insectos, entre la necesidad y el placer gastronómico. Etnofarmacia. 2001;20:150-161

[30] Sánchez-Muros MJ, Barroso FG, Manzano-Agugliaro F. Insect meal as renewable source of food for animal feeding: A review. Journal of Cleaner Production. 2014;65:16-27

[31] Sancho D, Gil MDJA, Sánchez LDRF. Insectos y alimentación. Larvas de *Rhynchophorus palmarum L*, un alimento de los pobladores de la Amazonía Ecuatoriana. Entomotropica. 2015;30(14):135-149

[32] Sänchez P, Jaffé K, Hevia P. Consumo de insectos: Alternative proteica del Neotrópico. Boletín de Entomología Venezolana. 1997;12(1):125-127

[33] Domínguez E, Peters WL, Peters JG, Savage HM. The imago of Simothraulopsis Demoulin with a redescription of the nymph (*Ephemeroptera*: *Leptophlebiidae*: *Atalophlebiinae*). Aquatic Insects. 1997;19(3):141-150

[34] De Foliart GR. Insects as human food: Discusses some nutritional and economic aspects. Crop Protection. 1992;**11**(5):395-399

[35] De Foliart GR. The human use of insects as food and animal feed. Bulletin of the Entomological Society of America. 1989;**35**:22-35

[36] Reid WV. How many species will there be. In: Tropical Deforestation and Species Extinction. 1992. pp. 55-57

[37] FAO. Los insectos comestibles: perspectivas de futuro de la seguridad alimentaria y la alimentación. Roma: FAO; 2012

[38] Araujo-Gutiérrez GP. Los insectos: una materia prima promisoria ontra la hambruna. Revista Lasallista de Investigación. 2005;**2**(1):33-37

[39] Sancho D. *Rhynchopohorus palmarum* (*Coleoptera: Curculionidae*) en la Amazonía, un insecto en la alimentación tradicional de las comunidades nativas. Revista Amazónica Ciencia y Tecnología. 2012;**1**(1):51-57

[40] Van Huis A. Potential of insects as food and feed in assuring food security. Annual Review of Entomology. 2013;**58**:563-583

[41] Van Huis A, Van Itterbeeck J, Klunder H, Mertens E, Halloran A, Muir G, et al. Edible Insects: Future Prospects for Food and Feed Security (No. 171). Rome: Food and Agriculture Organization of the United Nations; 2013

[42] Dossey AT, Méndez-Gutiérrez IR. Los insectos como una fuente de proteína limpia y sustentable para el futuro. Entomología Mexicana. 2014;**1**:1039-1044

[43] Cerda H, Martínez R, Briceño N, Pizzoferrato L, Paoletti MG. Cría, análisis nutricional o sensorial del picudo del cocotero (*Rhynchophorum palmarum*), insecto de la dieta tradicional indígena amazónica. Ecotropicos. 1999;**12**(1):25-32

[44] Herrera MC, Rivero-Martínez J, Melo V. Consumo de ortópteros alrededor del mundo. In: Navarrete-Heredia JL, Castaños-Meneses G, Quiroz-Rocha GA, editors. Facetas de la Ciencia. Ensayos Sobre Entomología Cultural. Guadalajara: Universidad de Guadalajara; 2011. pp. 85-86

[45] Brusca RC, Brusca GJ. Invertebrados. Madrid: McGraw-Hill; 2005. ISBN 9788448602468

[46] Paoletti M et al. Importance of leaf-and-litter-feeding invertebrates as source of animal protein for the Amazonian Amerindians. Proceedings of the Royal Society London B. 2000;**267**:2247-2252

[47] Jara F. La miel y el aguijón. Taxonomía zoológica y etnobiología como elementos en la definición de las nociones de género entre los Andoke (Amazonía colombiana). Journal de la Société des Americanistes. 1996;**82**:209-258

[48] Posey DA. Enthomological considerations in Southeastern aboriginal demography. Etnohistory. 1976;**23**(2):147-160

[49] Beckerman S. The use of palms by the Bari Indians of the Maracaibo Basin. Principles. 1977;**21**:143-154

[50] Sánchez P, Jaffe K, Hevia P. Consumo de insectos: alternativa proteíca del Neotrópico. Boletín de Entomología Venezolana. 1997;**12**(1):125-127

[51] Myers N. Homo insectivorus. Ciencia ilustrada. 1983:86-88

[52] Watanapongsiri A. A revision of the genera *Rhynchophorus* and

Dynamis. Department of Agriculture Science Bulletin Bangkok (Thailand). 1966;**1**:1-185

[53] Jaffé K, Sánchez P, Cerda H, Hernández JV, Jaffé R, Urdaneta N, et al. Chemical ecology of the palm weevil *Rhynchophorus palmarum* (L.) (*Coleoptera*: *Curculionidae*): Attraction to host plants and to a male-produced aggregation pheromone. Journal of Chemical Ecology. 1993;**19**(8):1703-1720

[54] Sánchez PA, Cerda H. The *Rhynchophorus palmarum* (L.) (Coleoptera: Curculionidae)-*Bursaphelenchus cocophilus* (Cobb) (Tylenchidae: Aphelenchoididae) complex in Palmeras. Boletin de Entomologia Venezolana. 1993;**8**(1):1-18

[55] Aldana de la Torre RC, Aldana de la Torre JA, Moya OM, Bustillo Pardey AF. Anillo rojo en palma de aceite. In: Boletín Técnico No. 36. Bogotá: Cenipalma; 2015

[56] Mexzón-Vargas RG, Chinchilla CM, Castrillo G, Salamanca D. Biología y hábitos de *Rhynchophorus palmarum L.* asociado a la palma aceitera en Costa Rica. ASD Oil Palm Papers. 1994;(8):14-12

[57] Griffith R. Red ring disease of coconut palm. Plant Disease. 1987;**71**(2):193-196

[58] Aldana de la Torre RC. Manejo del Picudo *Rhynchophorus palmarum*. Bogotá: Ministerio de Agricultura y Desarrollo Rural-ICA; 2011

[59] Genty P, De Chenon RD, Morin JP, Korytkowski CA. Oil palm pests in Latin America. Oléagineux. 1978;**33**(7):325-419

[60] González P, García U. Ciclo biológico de *Rhynchophorus palmarum* (*Coeóptera*: *Curcunlionidae*) sobre *Washington robusta* en laboratorio. Revista Peruana Entom. 1992;**35**:60-62

[61] Pérez D, Innacore. Aspectos de la biología del *Rhynchophorus palmarum* L. (*Coléptera*: *Curculionidae*) en el pijuayo (*Bactris gasipaes HBK*) (*Arecaceae*) en la Amazonía peruana. Revista Peruana Entomologia. 2006;**45**:138-140

[62] Sánchez P, Jaffe K, Hevia P. Consumo de insectos: alternativa proteica del Neotrópico. Boletín de Entomología Venezolana. 1993;**12**(1):125-127

[63] González N, Camino L. Biología y hábitos de *Rhynchophorus palmarum* en Contolpa, Tabasco (México), *Coleoptera*: *Curculionidae*. Revista Entomológica Mexicana. 1974;**28**:13-19

[64] Hagley EA. On the life history and habits of the palm weevil, *Rhynchophorus palmarum*. Annals of the Entomological Society of America. 1965;**58**(1):22-28

[65] Cerda H, Martínez R, Briceño N, Pizzoferrato L, Paoletti MG. Palm worm (*Rhynchophorus palmarum*). Traditional food in Amazonas, Venezuela. Nutritional composition, small scale production and tourist palability, production. Ecology of Food and Nutrition. 2001;**40**(1):13-32

[66] Barragán A, Carpio C. Plantas como alimentos de invertebrados útiles. In: Enciclopedia de las Plantas Útiles del Ecuador. Herbario QCA de la Escuela de Ciencias Biológicas de la Pontificia Universidad Católica del Ecuador & Herbario AAU del Departamento de Ciencias Biológicas de la Universidad de Aarhus. Quito; 2008. pp. 76-79

[67] Cartay R. La mesa amazónica peruana. Ingredientes, corpus, símbolos. Lima: Universidad San Martín de Porres; 2016

[68] Chinchilla C, Escobar R. El anillo rojo y otras enfermedades de la palma aceitera en Centro y Suramérica. ASD Oil Palm Papers. 2007;**30**:1-27

[69] Chinchilla C, Menjivar R, Arias E. Picudo de la palma y enfermedad del anillo rojo/hoja pequeña en una plantación comercial en Honduras. Turrialba. 1990;**40**(4):471-477

[70] Dufour D. Insects and food: A case study from the Northwest Amazon. American Anthropologist. 1987;**89**(2):383-397

[71] Vargas GE et al. Valor nutricional de la larva de *Rhynchophorus palmarum L.*: comida tradicional de la Amazonía peruana. Revista de la Sociedad Química del Perú. 2013;**79**(1):64-70

[72] Smith NJH. The Enchanted Amazon Rain Forest: Stories from Vanishing World. Gainesville: The University Press of Florida; 1996

[73] Chirif A. Diccionario Amazónico. Región Loreto: Lima: CAAAP-Lluvia Editores; 2016

[74] Brewer-Carías C. Desnudo en la selva: supervivencia y subsistencia. Caracas: CBC; 2013

[75] Krajick K. A swarm of tasty threats. The Food Insects Newsletter. 1994;**7**(2):3-4

[76] Bukkens SGF. Insects in the human diet. Nutritional aspects. In: Paoletti MG, editor. Ecological Implications of Minilivestock: Potential of Insects, Rodent, Frogs, and Snails. USA: Science Publishers Inc; 2005

[77] Ramos-Elourdoy J. Insects: A hopeful food source. In: Paoletti MG, editor. Ecological Implications of Minilivestocks: Potential of Insects, Rodents, Frogs and Snails. USA: Science Publishers Inc.; 2005. pp. 263-291

[78] Váldez C, Untiveros G. Extracción y caracterización de las larvas del Tenebro molitor. Revista de la Sociedad Química del Perú. 2010;**76**(4):407-414

[79] Sutton MQ. Insects as food: Aboriginal enthomofagy in the Great Basin. In: Anthropological Papers, 33. Menlo Park, California: Ballena Press; 1998

[80] Ramos-Elorduy J, Pino JM, Márquez C, Rincón F, Alvarado M, Escamilla E, et al. Protein content of some edible insects in Mexico. Journal of Ethnobiology. 1984;**4**(6)

[81] Costa-Neto EM. Insetos como fontes de alimentos para o homem. Valoración de recursos considerados repugnantes. Interciencia. 2003;**28**(3):136-146

[82] Ramos-Elourdoy J et al. Estudio comparativo del valor nutritivo de varios coleóptera comestibles de México y *Pachymerus nucleorum* (Bruchidae) de Brasil. Interciencia. 2006;**31**(7):512-516

[83] Ortiz-Quijano R. Conocimiento, uso y manejo de plantas silvestres y cultivadas y otros recursos para la alimentación de los indígenas Yucuna del resguardo indígena Miriti-Pavana. In: Estrella E, Crespo A, editors. Salud y población indígena de la Amazonia. Memorias del I Simposio Salud y Población Indígena de la Amazonia. Vol. 1. Quito: OTCA, Comisión de Comunidades Europeas, Museo Nacional de Medicina del Ecuador; 1993. pp. 3-28

[84] Arango-Gutiérrez GP. Los insectos: una materia prima alimenticia promisoria contra la hambruna. Revista Lasallista de Investigación. 2005;**2**(1)

[85] Miller PL. Dragonflies. Naturalist's Handbook 7. Cambridge: Cambridge University Press; 1987

[86] Wahlqvist ML, Lee M. Regional food culture and development. Asia Pacific Journal of Clinical Nutrition. 2007;**16**(Suppl. 1):2-7

[87] Acuña-Cors AM. Etología de insectos comestibles y su manejo tradicional por la comunidad indígena

de Los Reyes Metzontla, Municipio de Zapotitlán Salinas, Puebla. Tesis de Maestría en Ciencias; 2010

[88] Macera P, Casanto E. La Cocina mágica Ashaninka. Lima: Universidad San Martín de Porres; 2011

[89] Gerique A. An introduction to ethnoecology and ethnobotany: Theory and methods. Integrative Assesssment and Planning Methods for Sustainable Agroforestry in Humid Semiarid Regions. Senkenbergstr: University of Giessen; 2006

[90] Pino-Moreno JM et al. Análisis comparativo del valor nutricional de la Cuetla (*Arsenura armida C*, 1779) (*Lepidóptera*: *Saturniidae*), con algunos alimentos convencionales. Entomología Mexicana. 2015;**2**:744-748

[91] Kuhnlein HV. Micronutrients, nutrition and traditional food systems of indigenous peoples. Fnajana. 2003;**32**(2003):33-39

[92] Ramos-Elourdoy J. La etnoentomología en la alimentación, la medicina y el reciclaje. In: Llorente JB et al., editors. Biodiversidad, taxonomía y biogeografía de artrópodos de México: hacia una síntesis de su conocimiento. Vol. 4. México: UNAM; 2004. pp. 329-413

Chapter 7

The Potential of Insect Farming to Increase Food Security

Flora Dickie, Monami Miyamoto and C. Matilda (Tilly) Collins

Abstract

Insect protein production through 'mini-livestock farming' has enormous potential to reduce the level of undernutrition in critical areas across the world. Sustainable insect farming could contribute substantially to increased food security, most especially in areas susceptible to environmental stochasticity. Entomophagy has long been acknowledged as an underutilised strategy to address issues of food security. This chapter reviews and provides a synthesis of the literature surrounding the potential of insect farming to alleviate food security while promoting food sovereignty and integrating social acceptability. These are immediate and current problems of food security and nutrition that must be solved to meet the UNDP Sustainable Development Goals.

Keywords: climate change, sustainability, entomophagy, insectivory, acceptance

1. Introduction

Entomophagy is prevalent in many regions, and ~1500–2000 species of insects and other invertebrates are consumed by 3000 ethnic groups across 113 countries in Asia, Australia and Central and South America [1]. Africa, where more than 500 species are consumed daily, is a hotspot of edible insect biodiversity [2, 3]. In Thailand, entomophagy has spread to the south from the north-east as people migrate towards city centres. It has become so popular that >150 species are sold in the markets of Bangkok [4]. The most common edible insects are moths, cicadas, beetles, mealworms, flies, grasshoppers and ants [5]. Although human insectivory is an ancient practice and 80% of the world's population consumes insects, it is relatively uncommon in contemporary Western culture. In many regions that have traditionally eaten insects, the practice is declining due to globalisation, and their consumption has decreased over the last decade as agriculture and living standards change, and the availability of wild-caught insects has decreased [6–8].

This chapter reviews and provides an accessible synthesis of the literature surrounding the potential of insects to alleviate food security while promoting food sovereignty and integrating social acceptability. These are immediate and current problems of food security and nutrition that must be solved to meet the Sustainable Development Goals [3, 9].

2. Food insecurity

Food insecurity is created when food is unavailable, unaffordable, unevenly distributed or unsafe to eat. Inefficiencies in the current food production system generate inconsistencies between the demand and supply of food resources, which is exacerbated by the diminution of pastures and increasing demand for food. Thirty percent of land is already used for agriculture, but 70% of this is used for macro-livestock production, an industry which consumes 77 million tonnes of plant protein only to produce 58 million tonnes of animal protein per year. This animal protein is not evenly distributed across the globe, as the average person in a 'developed' country consumes 40 g more protein a day than the average person in a 'developing' country [10]. The demand for affordable and sustainable protein is high, while animal protein is becoming more expensive and less accessible in some regions, especially in Africa [11].

To ensure food access and to alleviate poverty, there is a particular need for investment into Africa's agricultural potential as this continent will soon account for 50% of the world's population growth. Currently, Africa has 25% of all under-nourished people worldwide, and the income gap between rural and urban areas drives rapid urbanisation; this is decreasing the agricultural workforce [12, 13]. With substantial food insecurity and rising food prices, one in six people dies from malnutrition and hunger, and more than 1 billion people are undernourished, triggering 1/3 of the child disease burden [10, 14]. Effects are worse in the populations that already have high rates of malnutrition, such as Zambia, where chronic undernutrition is 45% and causes 52% of deaths in the population under the age of 5. Over 800 million people are thought to have a food energy deficit average of >80 kcal/day/person [3, 15].

The prospect of global food shortage grows as the world's population is esti-mated to increase to 9 billion by 2050. The conventional meat production system will not be able to respond sufficiently to the increase in demand. *Per capita*, meat consumption is expected to increase by 9% in high-income countries by 2030, and the increase in world crop prices will increase the price of meat by 18–21% [16]. Systems with a low carbon footprint must be promoted according to the economic and cultural restraints of the region by modifying animal feed from soy meal to locally sourced feed [17]. Any expansion of agricultural land must be mitigated to reduce losses in natural ecosystems. Therefore, our increasing population will need to be fed from the same area of land available now [18].

Climate change is also a growing threat to global food security as this is reducing the area of land available to agriculture [10], and future cereal yields are predicted to decrease, especially in low-latitude areas. The poorest countries will suffer the worst consequences of climate change, which will increase both malnutrition and poverty. To prevent future undernutrition and to decrease current levels, food access and socioeconomic conditions must improve globally [14]. With this climate change-driven prediction of reduced agricultural yields in most countries given current crop practices and varieties, it is therefore necessary to increase the diver-sity and sustainability of crop supply so that food insecurity is not exacerbated [15].

3. Nutritional potential of edible insects

In general, insects have a higher quality of nutrition than macro-livestock in terms of protein, lipids, carbohydrates and vitamins [10]. Insects have high crude protein levels of 40–75%, contain all essential amino acids, are rich in fatty acids and have a high proportion of dietary fibre, and it has been further suggested that there are health benefits from eating chitin through enhancement of gut flora and antibiotic properties, though it is not known how insect fibre specifically affects human health [19].

In a study of the calorific value of 94 insect species, 50% were higher than soybeans, 87% higher than maize, 63% higher than beef and 70% higher than fish [10]. The composition of omega-3 and omega-6 fatty acids in mealworms is comparable to that of fish, and other insects with ideal fatty acid ratios are house crickets, short-tailed crickets, Bombay locusts and scarab beetles [20]. Some insect species have micronutrients not found in some conventional animal proteins, such as riboflavin in termites and high concentrations of thiamine in silk moth larvae (224.7% daily human requirement) and palm weevils (201.3%) compared to chicken (5.4%). Mealworms have a higher content of protein (all essential amino acids), calcium, vitamin C, thiamine, vitamin A and riboflavin per kg than beef. Although the nutritional content of many insects is well-described in the literature, there is a variation depending on diet, sex, life stage, origin and environmental factors, and the realised nutritional content also depends on preparation and cooking [21–23].

Insect consumption has the potential to reduce hunger on a global scale as they are nutrient dense as well as calorie dense. A calorie deficit of 1500 kcal/day could be addressed by rearing 1 kg/day of crickets in 10 m^2 while also providing the recommended daily amount of lysine, methionine, cysteine, tryptophan, zinc and vitamin B$_{12}$. Not only do insects provide calories and nutrients, but they are also cost-effective, easily grown and can be environmentally sustainable when incorporated into a circular production system using organic side streams.

4. The rise of insect farming

Until the end of the twentieth century, the most common way to collect insects worldwide was by wild harvest (circa 90%), and the tradition of collecting and eating insects from the wild is seen in many cultures. Though seasonality limits consistent availability, traditional regulation patterns can mitigate this and maintain locally sustainable sources [24, 25]. Wild catch is declining in many areas with many factors contributing to this including land conversion, overexploitation and urbanisation [7]. With insects acknowledged to be key to the delivery of many ecosystem services, their conservation in natural ecosystems is now paramount [26, 27]. In response, the farming of edible insects is now rising from being only a minor component of the market and should be promoted to improve quality and supply as well as to limit the environmental impacts of wild harvesting [11, 28].

No matter the scale of insect farming, the economic benefits boost food security in terms of availability and accessibility and at the same time improve dietary quality and contribute to both gender equity and livelihoods. At the community scale, more than 20,000 small farmers in Thailand profitably produce crickets; in Laos, the majority of insect vendors are illiterate females who may earn c$5/day; in Uganda and Kenya, the Flying Food Project supports expansion of small-scale farms into local and greater value chain markets [20, 29, 30]. By integrating mini-livestock farming into current agricultural systems, the access to edible insects could be improved and simultaneously provide co-benefits such as female employment and a high-grade compost contribution to the enhancement of soil fertility [28]. Harvesting insects as a by-product of another industry also has substantial potential but needs more widespread implementation and cultural assimilation. For example, domesticated silkworms for the textile industry can be eaten in the

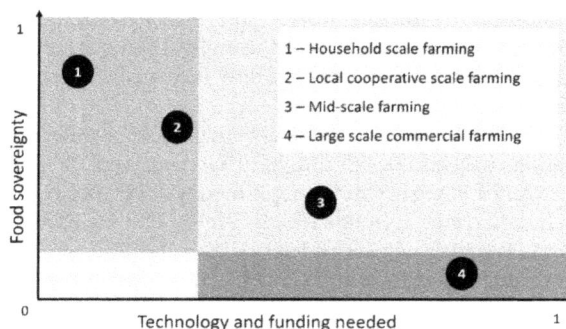

Figure 1.
Trade-offs in the scale of production needed to maximise food sovereignty relative to the technology and initial funding needed. X axis: 0 = none needed, 1 = high setup costs needed. Y axis: 0 = no food sovereignty, 1 = complete food sovereignty.

pupa stage, and palm weevils reared on felled palm trees could be moved into more formal production [15]. Insect farming is now moving into western markets and developing technologically refined production systems. The French company Ynsect has raised \$175 M for expansion, and the USA edible insect market is predicted to increase by 43% in the coming 5 years [31, 32]. There are different costs and benefits at all scales (**Figure 1**), though all may have an important place in future food security.

5. The environmental advantage of insect farming

In general, insects have a lower consumption of energy and resources than conventional animal livestock. Insects are poikilothermic, so they expend less energy, are more efficient in transforming phytomass into zoomass and have higher fecundity and growth rates and a higher rate of matter assimilation. On average, an insect only needs 2 g of food per gramme of weight gained, whereas a cow needs 8 g of food. Not only is the efficiency of insect production higher because of the feed conversion ratio (**Table 1**) but also because the edible portion of insects is higher as crickets can be eaten whole, but we only eat 40% of a cow, 58% of a chicken and 55% of a pig [8, 10, 33].

Edible insects are an environmentally attractive alternative to conventional livestock because they require less feed and water; they produce lower levels of greenhouse gases and can be raised in small spaces. Worldwide, livestock contributes to 18% of greenhouse gas emissions, which, in light of global warming and climate change, favours the less resource-intensive insect production which emits fewer greenhouse gases by a factor of 100 [3, 28].

Insects can be a renewable food source in the future as many edible species can consume agricultural and food waste or culinary by-products, but there remain important research gaps in understanding the effects of variable feedstocks as most case studies use high-grade feed [10, 15, 28]. Such organic side streams could be used to reduce the environmental impact of insect farming while simultaneously creating a novel, circular waste-processing income. Throughout the world, 1/3 of all food is wasted, and household food waste is 70% of the post-farm total. If food waste was its own country, it would be the third largest emitter of greenhouse gases after the USA and China [30]. Food waste is expected to increase in the future

	Cricket	Poultry	Pork	Beef
Feed conversion ratio (kg feed: kg live weight)	1.7	2.5	5	10
Edible portion (%)	80	55	55	40
Feed (kg: kg edible weight)	2.1	4.5	9.1	25

Table 1.
Efficiencies of production of conventional meat and crickets [17].

with a continually growing and increasingly urbanised global population adopting 'modern' lifestyles.

It is challenging and wasteful to commercialise traditional composting of multiple waste streams on a large scale, but waste can be fed directly to insects to convert low-value biomass into higher-value insect mass. By valorising waste as feed, it may mitigate the impact of the food industry. Some fly (Diptera) species are known to be able to convert agricultural manure into body mass and reduce the waste dry matter by 58%. For food waste the conversion is as high as 95% leaving the remainder as a high-grade soil improver [30, 33].

6. Acceptability of eating insects as animal protein

The feasibility of promoting edible insect farming as sustainable protein depends on social acceptance, as the benefits cannot be realised if people do not choose to eat insects. The understanding of current perceptions, which often depend on class, location, gender and age, is essential to any market development. In some locations, newly urbanised citizens view insects as pests or as poor person's food [7]. Although in this particular case, acceptance does depend on the insect itself, as there is an inferiority complex associated with wild harvesting of insects. In the Western world, insects are largely unfamiliar and mostly viewed as holiday novelty or 'yuk'; thus, awareness of local taboos, cultural preferences and the population's exposure to insects as food are crucial for the successful promotion of insect farming for food [3, 15, 34].

In many urban and developed populations, a central issue is food neophobia, but after taking the first step in trying an insect, continued exposure correlates with increased acceptance. Processed insect products such as cookies, snack bars or powders further normalise the protein source [34, 35]. Conventional meat has a special status in society, both culturally and structurally in meals, so a sustainable culinary culture must be promoted in order to associate insect protein with pleasurable food [17].

There are also risk considerations with the dissemination of novel foods and novel production pathways. Possible effects of prolonged insect consumption are nutrient malabsorption, growth alteration, allergy risk and contamination, and more research is needed into the digestion and absorption of insects in the human body [36]. Intensive insect farming runs risks of microbial infestation, parasites and pesticides. Preventative approaches, such as probiotics, transgenerational immune priming or heat treatment, and measured responses such as those advocated by Integrated Crop Management (ICM) will develop with the industry [20, 37]. There are other limitations in the lack of protocols in storage and decontamination, and although international regulation is underway, these ancient foods are currently classified the EU as novel foods [38].

7. Conclusion

The issue of food security is multi-faceted, and each country's solution will be different. Tackling food security requires responses that are both innovative and culturally appropriate. Farming insect livestock has the potential to alleviate food insecurity while promoting food sovereignty, especially if it is integrated with social acceptability in mind. Engagement of all stakeholders on the production and consumption sides and continued support for and from them will be vital for the success of its implementation. Commercial farming is growing across Europe and the North American continent, though a question yet to be answered at a wider scale is how edible insect farming can be increased and deployed in a way that benefits all parties, including especially the most vulnerable. We have overviewed the field and hope that this synthesis of much important work along with the exemplar production model of **Figure 2** can provide encouragement and compact information to those seeking to evaluate the future of farmed insect production.

There is currently too little research available on the integration of insect farming with existing agricultural systems, and future solutions require the coordination of international, national and legal frameworks. With this in place, the future food revolution will be more able to directly benefit the poor and be environmentally sustainable [39].

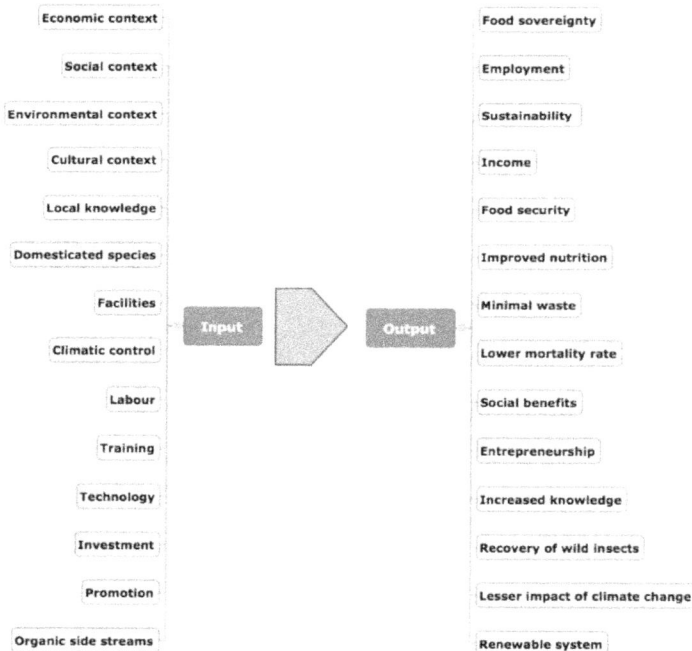

Economic context		Food sovereignty
Social context		Employment
Environmental context		Sustainability
Cultural context		Income
Local knowledge		Food security
Domesticated species		Improved nutrition
Facilities	Input → Output	Minimal waste
Climatic control		Lower mortality rate
Labour		Social benefits
Training		Entrepreneurship
Technology		Increased knowledge
Investment		Recovery of wild insects
Promotion		Lesser impact of climate change
Organic side streams		Renewable system

Figure 2.
Idealised schematic of the inputs and outputs of a sustainable production model for insect farming.

Acknowledgements

The authors wish to thank Harry McDade who contributed to the discussions on this topic. Thanks also go to the many who have written so passionately on this topic and to the inspiring Arnold van Huis; may these efforts eventually bear fruit, or larvae. Particular thoughts go to Dr. Marianne Schockley of the University of Georgia, Athens, GA, who advocated so ably and enthusiastically for Entomophagy in the USA.

Conflict of interest

The authors declare no conflict of interest.

Author details

Flora Dickie[1], Monami Miyamoto[1] and C. Matilda (Tilly) Collins[2]*

1 Department of Life Sciences, Imperial College London, London, United Kingdom

2 Centre for Environmental Policy, Imperial College London, London, United Kingdom

*Address all correspondence to: t.collins@imperial.ac.uk

IntechOpen

References

[1] MacEvilly C. Bugs in the system. Nutrition Bulletin. 2000;**25**(4):267-268

[2] Kelemu S, Niassy S, Torto B, Fiaboe K, Affognon H, Tonnang H, et al. African edible insects for food and feed: Inventory, diversity, commonalities and contribution to food security. Journal of Insects as Food and Feed. 2015;**1**(2):103-119

[3] Stull VJ, Wamulume M, Mwalukanga MI, Banda A, Bergmans RS, Bell MM. "We like insects here": Entomophagy and society in a Zambian village. Agriculture and Human Values. 2018;**35**(4):867-883

[4] Yhoung-Aree J, Viwatpanich K. Edible insects in the Lao PDR, Myanmar, Thailand and Vietnam. In: Paoletti MG, editor. Ecological implications of minilivestock: Potential of insects, rodents, frogs and snails. Enfield, NH, USA: Science Publisher Inc; 2005. pp. 415-440

[5] Ramos-Elorduy J. Anthropo-entomophagy: Cultures, evolution and sustainability. Entomological Research. 2009;**39**:271-288

[6] Belluco S, Losasso C, Maggioletti M, Alonzi CC, Paoletti MG, Ricci A. Edible insects in a food safety and nutritional perspective: A critical review. Comprehensive Reviews in Food Science and Food Safety. 2013;**12**(3):296-313

[7] Looy H, Dunkel FV, Wood JR. How then shall we eat? Insect-eating attitudes and sustainable foodways. Agriculture and Human Values. 2014;**31**(1):131-141

[8] Vogel G. For more protein, filet of cricket. Science. 2010;**327**(5967):881

[9] Tomberlin JK, Zheng L, van Huis A. Insects to feed the world conference 2018. Journal of Insects as Food and Feed. 2018;**4**(2):75-76

[10] Premalatha M, Abbasi T, Abbasi T, Abbasi SA. Energy-efficient food production to reduce global warming and ecodegradation: The use of edible insects. Renewable and Sustainable Energy Reviews. 2011;**15**:4357-4360

[11] Raheem D, Carrascosa C, Oluwole OB, Nieuwland M, Saraiva A, Millán R, et al. Traditional consumption of and rearing edible insects in Africa, Asia and Europe. Critical Reviews in Food Science and Nutrition. 2018;**15**:1-20

[12] Sasson A. Food security for Africa: An urgent global challenge. Agriculture and Food Security. 2012;**1**(2)

[13] Parnell S, Walawege R. Sub-Saharan African urbanisation and global environmental change. Global Environmental Change. 2011;**21**(suppl 1): 12-20

[14] Lloyd SJ, Sari Kovats R, Chalabi Z. Climate change, crop yields, and undernutrition: Development of a model to quantify the impact of climate scenarios on child undernutrition. Environmental Health Perspectives. 2011;**119**(12):1817-1823

[15] Laar A, Kotoh A, Parker M, Milani P, Tawiah C, Soor S, et al. An exploration of edible palm weevil larvae (Akokono) as a source of nutrition and livelihood: Perspectives from Ghanaian stakeholders. Food and Nutrition Bulletin. 2017;**38**(4):455-467

[16] van Huis A. Potential of insects as food and feed in assuring food security. Annual Review of Entomology. 2013;**58**(1):563-583

[17] van der Spiegel M, Noordam MY, van der Fels-Klerx HJ. Safety of novel

protein sources (insects, microalgae, seaweed, duckweed, and rapeseed) and legislative aspects for their application in food and feed production. Comprehensive Reviews in Food Science and Food Safety. 2013;**12**:662-678

[18] Oonincx DGAB, de Boer IJM. Environmental impact of the production of mealworms as a protein source for humans: A life cycle assessment. PLoS ONE. 2012;7:12

[19] Ozimek L, Sauer WC, Kozikowski V, Ryan JK, Jørgensen H, Jelen P. Nutritive value of protein extracted from honey bees. Journal of Food Science. 1985;**50**(5):1327-1329

[20] Barennes H, Phimmasane M, Rajaonarivo C. Insect consumption to address undernutrition, a national survey on the prevalence of insect consumption among adults and vendors in Laos. PLoS ONE. 2015;**10**(8)

[21] Payne CLR, Scarborough P, Rayner M, Nonaka K. Are edible insects more or less "healthy" than commonly consumed meats? A comparison using two nutrient profiling models developed to combat over- and undernutrition. European Journal of Clinical Nutrition. 2016;**70**(3):285-291

[22] van Huis A, Oonincx DGAB. The environmental sustainability of insects as food and feed: A review. Agronomy for Sustainable Development. 2017;**35**(7):1-14

[23] Banjo A, Lawal O, Sononga E. The nutritional value of fourteen species of edible insects in southwestern Nigeria. African Journal of Biotechnology. 2006;5:298-301

[24] Illgner P, Nel E. The geography of edible insects in sub-Saharan Africa: A study of the mopane caterpillar. The Geographical Journal. 2000;**166**(4):336-351

[25] Mbata KJ, Chidumayo EN, Lwatula CM. Traditional regulation of edible caterpillar exploitation in the Kopa area of Mpika district in northern Zambia. Journal of Insect Conservation. 2002;6(115)

[26] Losey JE, Vaughn M. The economic value of ecological services provided by insects. Bioscience. 2006;**56**(4):311

[27] Sánchez-Bayo F, Wyckhuys KAG. Worldwide decline of the entomofauna: A review of its drivers. Biological Conservation. 2019;**232**:8-27

[28] Nadeau L, Nadeau I, Franklin F, Dunkel F. The potential for entomophagy to address undernutrition. Ecology of Food and Nutrition. 2015;**54**(3):200-208

[29] Halloran A, Vantomme P, Hanboonsong Y, Ekesi S. Regulating edible insects: The challenge of addressing food security, nature conservation, and the erosion of traditional food culture. Food Security. 2015;7(3):739-746

[30] Entomics. Entomics [Internet]. Available from: www.entomics.com

[31] Ynsect [Internet]. 2019. Available from: http://www.ynsect.com/en/

[32] Ahuja K, Deb S. Edible insects: Market size by product, by application, industry analysis report, regional outlook, application potential, price trends, competitive market share and forecast, 2018-2024. Delaware, USA: Global Market Insights; 2018

[33] van Huis A, Klunder JVIH, Merten E, Halloran A, Vantomme P. Edible Insects. Future Prospects for Food and Feed Security. Rome: Food and Agriculture Organization of the United Nations; 2013

[34] Collins CM, Vaskou P, Kountouris Y. Insect food products in the Western

world: Assessing the potential of a new 'green' market. Annals of the Entomological Society of America. 2019. IN PRESS

[35] Hartmann C, Siegrist M. Becoming an insectivore: Results of an experiment. Food Quality and Preference. 2016;**51**:118-122

[36] Testa M, Stillo M, Maffei G, Andriolo V, Gardois P, Zotti CM. Ugly but tasty: A systematic review of possible human and animal health risks related to entomophagy. Critical Reviews in Food Science and Nutrition. 2017

[37] Grau T, Vilcinskas A, Joop G. Sustainable farming of the mealworm Tenebrio molitor for the production of food and feed. Zeitschrift fur Naturforschung: Section C Journal of Biosciences. 2017;**72**(9):337-349

[38] Finke MD, Rojo S, Roos N, van Huis A, Yen AL. The European food safety authority scientific opinion on a risk profile related to production and consumption of insects as food and feed. Journal of Insects as Food and Feed. 2015;**1**(4):245-247

[39] Conway G, Wilson K. One Billion Hungry. 1st Editio ed. Ithaca, N.Y.: Comstock Publ. Assoc; 2012